A Literature
Guide for IDENTIFYING
MUSHROOMS

A Literature Guide for IDENTIFYING MUSHROOMS

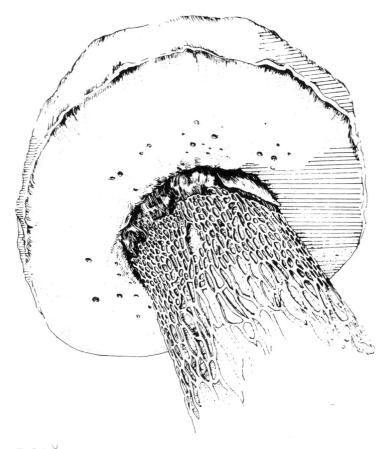

by ROY and
ANN ELIZABETH WATLING

MAD RIVER PRESS INC.

Published by Mad River Press, Inc.
Route 2, Box 151-B
Eureka, California 95501

Typeset and Printed by Eureka Printing Company, Inc.
Eureka, California 95501

ISBN 0-916-422-18-6

Copyright © 1980, Roy and Ann Elizabeth Watling

MRP books printed in the English language are distributed outside of North and South America by the Richmond Publishing Company, Orchard Road, Richmond, Surrey, England TW94PD.

TABLE OF CONTENTS

PREFACE

On my husband's return from Australia in 1974, I assisted him in preparing a list of the mongraphic treatments of agarics which he used routinely in his studies. This was prepared for the British Council as a programme whose function was to attempt to alleviate the apparent lack of a working knowledge of the literature on larger fungi in Australasia which was primarily due to the inavailability of publications, both texts and periodicals. The situation found in India in 1978 was even more acute, although there was evidence that the appropriate journals were to be found in some library in that vast country. Two lists have been prepared. The first list was originally from a file-index of reprints and was later expanded to include references to tropical and subtropical groups; this second list, like the first, was expanded to include other data and was distributed by the British Council to the Indian workers met by Roy in 1978.

Many people who had seen the first list indicated how useful it was and so my husband persuaded me, with his help, to expand the original lists to include many more references and to expand the entries to give some indication as to their contents and usefulness. This has been a bibliographic exercise and whenever any problems were encountered, Roy supplied the appropriate information from his notes.

As a student Roy found the bibliography at the end of Ernest Athearn Bessey's 1952 edition of *Morphology and Taxonomy of the Fungi* extremely useful and consulted it frequently not only during his college days but also many times thereafter. The bibliography was kept up-to-date for his own use and these additions were found to be invaluable when offering assistance to Margaret Holden during her production of the very useful compilation 'Literature for the Identification of British Fungi' in the *Bulletin of the British Mycological Society*, 1952. A recent and much more exclusive production was prepared by the Systematic Association in Britain and appears in Kerrich, Hawksworth and Sims' tabulation of books and articles useful in the identification of organisms including all groups of fungi found in N. Western Europe. (Key words to the Fauna and Flora of the British Isles and N.W. Europe. G.J. Kerrich, D.C. Hawksworth and R.W. Sims. Syst. Assoc., Special Vol. 9, 1978 Academic Press, London). It is hoped that this present text goes one stage further and fills a much needed gap in our library.

Ann Elizabeth Watling
Edinburgh, Feb. 1980

INTRODUCTION

A. HOW TO USE THE BOOK

The present book is basically arranged according to the families found in How to Identify Mushrooms, Vol. III. However, due to the wide implications of many of the publications of Rolf Singer, some of those by him are tabulated in the first chapter. In the second chapter the references within each family are arranged chronologically under genera and these are then alphabetically listed by author except when the contribution covers general aspects of the entire family. The third chapter lists contain references which cover species from particular areas of the world; some of these also occur in the main taxonomic arrangement (chapter 2) but many do not. Cross-referencing between these two parts of this book should always therefore be conducted to make full use of all the references listed; to make cross-referencing easier for the reader an index will be found at the end of this book. Many of the references appear in journals and their abbreviations follow the World List of Scientific Journals, or if they do not the way they differ is indicated. A complete list of these journals in full is given in chapter 5 and if a wholly mycological publication, then the address of the sponsor or society is also given unless indicated. (see pg. 110). Abstracting of journals is now a regular feature of the *Bulletin of the British Mycological Society* and *Bulletin Trimestrie de la Societe Mycologique de France* and consultation of these publications, or better still, membership to the Societies, will allow the present book to be brought up-to-date. *The Bibliography of the Systematic Mycology*, a bi-annual publication produced by the Commonwealth Mycological Institute, Kew, England, lists all the current titles of publications on systematic mycology.

It started in 1961; some of this literature cited is abstracted in *Rev. Plant Pathology (Formerly Rev. Applied Mycology)* from the same institute and can be seen there. When using a publication the bibliography should always be consulted, as in this way little known references can be located.

There is a whole range of popular books on fungi available on the market and they appear to be increasing monthly. It was not considered necessary to include these herein as most do not allow a definite identification of a collection. However, a selection of books which offer good illustrations is included in part 4. Also included in chapter 4 is a list of general texts which, although not allowing identification of individual collections, may be necessary for consultation at one time or another.

It is hoped that with all the ammunition herein a good working knowledge of the literature will be achieved. Once this major hurdle is removed, the real business of identifying one's collections can be attempted. Unfortunately our knowledge of the northern Slavic, Japanese and Chinese publications is somewhat limited and omissions in this case are because of our ignorance of the language. However, with the continued correspondence of Drs. Wasser, Urbonas, and Kalamees, Drs. Hongo and Sagara and Dr. Zang Mu, Peking, it is possible to suggest some texts which have been found useful in the past. The references have been taken up as far as possible to the end of 1979; some later articles have crept in, others have not as yet come to our attention.

B. KEY CHARACTERS

It is often found that once the material has been returned to the laboratory for identification, indeed before a full description has been attempted, changes have already taken place in the specimens. Some texts use the original unchanged characters as key characters, e.g. colour of young gills in *Cortinarius* and *Agaricus* and weeping of the gills in *Hebeloma*, and without them identification is made, if not impossible, such more difficult. Therefore when introducing each major genus (often associated with those segregate genera outlined in 'How to Identify Mushrooms' IV) in this book, the important yet sometimes ephemeral characters necessary in the identification are listed; it is hoped that these will then be pin-pointed each time a collection is made. This does not excuse the collectors from preparing a full description but it insures that the important characters are never overlooked and also makes the professional mycologist's job so much easier when the collection is sent to him for further examination. Species differentiation is generally on colour of the basidiome coupled with microscopic data but whereas the first may change the latter can always be obtained even from herbarium material provided the specimens have been dried well. One exception is perhaps to be found in *Coprinus* where autolysis (deliquescence) of the basidiome takes place. Therefore in *Coprinus* it is advisable to examine the veil to recognize the major veil-type and record the morphology of the marginal and facial cystidia as soon as possible after collecting (see pg. 18). In the text below, other than these particular characters necessary for identification of species of *Coprinus*, macroscopic characters are tabulated. It is assumed that every good agaricologist notes the plant-community and the identity of any suspected host, whether the host of a mycorrhizal relationship or whether supplying the organic substrate.

C. ABBREVIATIONS USED IN THE LITERATURE CITATIONS

abbrev.	= abbreviation
ad int.	= meantime
Beih.	= Beihefte or supplement
b/w	= black and white
contd.	= continued
f.	= forma
fasc.	= fascicle
grp.	= group
p.p.	= in part
q.v.	= refer to
s.l.	= sub lente, under the lens
s.m.	= sub microscopica, under the microscope
s. lato	= in the widest sense
s. stricto	= in the restricted sense
sect.	= section
sp.	= species, plural spp.
ssp.	= subspecies, plural sspp.
subg.	= subgenus
suppl.	= supplement
var.	= variety

I. GENERAL REFERENCES OF ROLF SINGER

This author has been so prolific that it is necessary to offer a summary of those publications covering subjects it is not possible to categorize under the generic or country headings tabulated below. A full list of publications can be found in *Sydowia Beih.* 8 (1979); 1 - 13.

Singer, R. (1975). *The Agaricales in Modern Taxonomy*, Third edition 912 pp. 84 Taf., Verlag Cramer, Vaduz. This is a much expanded edition of the publication under the same title published in 1962 also by Cramer (Weinheim, 915 pp. and 73 Taf.). Both editions are based on the article which appeared in *Lilloa* 22 (1949) 1-832 in 1951. In 1962 the keys to families and genera, and some lower taxa were published separately by Cramer.

This is a fundamental work unparalleled by any author, except that R. Kuhner in *Bull. Soc. Linn. Lyon.* is presenting as a series his own summary of the agarics; yet Kuhner's work is not complete. A supplement to Singer's 3rd edition is found in *Nova Hedw.* 26 (1975); 435-436. Keys are being produced by Rolf Singer in *Sydowia* to cover taxa, at and below the species-level, not available in 'Agaricales in Modern Taxonomy'. This is an on-going project. So far the following have been published.

A - *Boletus. Sydowia* 30 (1978); 192-279.

B (continued) - *Clitocybe. Sydowia* 31 (1979); 193-237.

Singer, R. (1948-1973). Diagnoses Fungorum novorum Agaricalium. I. *Sydowia* 2 (1948); 26-43; II. Ibid 15 (1961); 45-83; III. *Sydowia, Beih 7; 1-106.*

Singer, R. (1936-1940). Notes sur quelques Basidiomycetes. I. *Rev. Mycol.* 1(1936); 75-85: II. Ibid 1 (1936); 279-293. III. Ibid 2 (1937); 226-242; IV. Ibid 4 (1938); 187-199: V. Ibid 4 (1939); 65-72: VI. Ibid 5 (1940); 3-13.

Singer, R. (1945-1960). New and interesting species of Basidiomycetes. I. *Mycologia* 37 (1945); 425-429: II. *Pap. Mich. Acad. Sci.* 34 (1948); 103-118: III. *Sydowia* 4 (1950); 130-157: IV. *Mycologia* 47 (1955); 763-777: V.*Sydowia* 11 (1957); 141-272: VI. *Mycologia* 51 (1959); 375-400: VII. Ibid 51 (1960); 578-594.

Singer R. (1942-1961). Type studies on agarics. I. *Lloydia* 5 (1942); 97-135: II.Ibid 9 (1946); 114-131; III. *Lilloa* 25 (1952); 463-514: IV. *Sydowia* 15 (1961); 133-151.

Singer, R. (1942-1961). Type studies on Basidiomycetes. I. *Mycologia* 34 (1942); 64-93: II. Ibid 35 (1943); 142-163: III Ibid 39 (1947); 171-189: IV. *Lilloa* 23 (1952); 147-246: V. *Sydowia* 5 (1951); 445-475: VI. *Lilloa* 26 (1953); 57-159: VII. *Sydowia* 6 (1952)); 344-351: VIII. Ibid 9 (1955); 367-431: IV. Ibid 13 (1959); 235-238: X. *Persoonia* 2 (1961); 1-62.

Singer, R. (1944-1964). New Genera of fungi. I. *Mycologia* 36 (1944); 358-369: II. *Lloydia* 8 (1945); 139-145: III. *Mycologia* 39 (1947); 77-89: IV. Ibid 40 (1948); 139-144: V. Ibid 43 (1951); 598-604: VI. *Lilloa* 23 (1950); 2-5-258 (1952): VII. *Mycologia* 48 (1956); 719-727: VIII. Ibid 50 (1958); 103-110: IX. *Lloydia* 21 (1958); 45-47: X. *Sydowia* 11 (1958); 320-322: XI. Ibid 16 (1963); 260-262: XII. Ibid 17 (1964); 12-16: XIII. Ibid 17 (1964); 142-145. see also VIII *Persoonia* 2 (1962); 407-415.

Prof. Rolf Singer also contributed to the Herbette Symposium on 'The Species Concept in Hymenomycetes' where he delivered a paper entitled 'The species concept in Agaricales and its adaption to taxonomy.' As this paper and those of the other participants to the symposium colour the references which are laid-out below it was considered important to list those taking part. Many of the authors you will see time and time again in the references below; it would therefore always be useful when utilizing the reference-source to marry the author's monographic treatment with his species concept - and that of others! The full publication is obtainable from J. Cramer, Vaduz as *Bibliotheca Mycologica* 61 (1977); 1-444.

Participants in order of appearance at the Symposium were as follows:

R. Watling: An analysis of the taxonomic characters used in defining the species of the Bolbitiaceae.

H.D. Thiers: Species concepts in the boletes.

C. Bas: Species concept in *Amanita* sect. *Vaginatae*.

H.E. Bigelow: Differentiation of species in *Clitocybe*.

A.H. Smith: Speciation in *Lactarius*.

R. Kuhner: Apropos de la delimitation des especes dans les *Hygrophorus* Fries du sous-genre *Hygrocybe* Fries. Deux caracteristiques peu ou non utilisees.

K. Esser and P. Hoffmann: Genetic basis for speciation . . .

R. Blaich: Enzymes as an aid in taxonomy . . .

A. Bresinsky, O. Hilber and H.P. Molitoris: The genus *Pleurotus* as aid for understanding the concept of species in Basidiomycetes.

R.F.O. Kemp: Oidial homing and the taxonomy . . .

J. Boidin and F. Oberwinkler each presented a paper on Aphyllophorales

H. Romagnesi: Incidence des caracteres non morphologiques . . .

R.H. Petersen: Species concept in Higher Basidiomycetes; Taxonomy, Biology, and Nomenclature

R. Singer: The species concept in Agaricales . . .

II. REFERENCES ARRANGED IN TAXONOMIC FRAMEWORK

Genera with an asterisk * should also be sought on pg. 71 under spore-studies.

AGARICACEAE (AGARICEAE)

Agaricus (syn. Psalliota)

Characters. 1. Staining of pileus and flesh of stipe-base. 2. Complexity of annulus. 3. Colour of young gills. 4. Morphology of stipe-base. 5. Scaliness and size of pileus. 6. Reaction with concentrated sulphuric acid and with conc. caustic potash. 7. Colour change in Schaeffer's reaction.

Hotson, J.W. and Stuntz, D.E. (1938). The genus *Agaricus* in western Washington. *Mycologia* 30; 204-234. Twenty-four species are listed and keyed-out; one new species is described and only one species fully documented. Good discussional information offered.

This paper is profusely illustrated with numerous b/w photographs and a single plate of line-drawings.

Mendoza, J.M. and Simeona, L-P. (1940). A revision of the genus *Psalliota* in the Phillippines. *Philipp. J. Sci.* 72; 337-345. A key and descriptions of thirteen species are given. Three new combinations are proposed. The paper is supported by eight b/w photographs.

Smith, A.H. (1940). Studies on the genus *Agaricus. Pap. Mich. Acad. Sci.* 25; 107-138. Forty-three species, arranged in major stirps based on European species, including two new are considered; two new varieties are described. Illustrated with line-drawings and b/w photographs. Extensive descriptions of the majority are supplied. Sporograms of thirteen additional tropical American species are also given although these are not supported by descriptive data.

Moeller, E.H. (1950-52). Danish *Psalliota* species. *Friesia* 4; 1-60 and f; 135-242. A standard work in two parts (*Rubescentes* and *Flavescentes*) giving full descriptions of most West European taxa. Several new taxa are described. Illustrated with line- drawings, b/w photographs and coloured icones, the last particularly of the new taxa.

Pilat, A. (1951). The Bohemian species of the genus *Agaricus. Sb. nar. Mus. Prazie* 78; 1- 142. A classic treatment important outside the geographical area indicated by the title. see Bohus below.

Heinemann, P. (1956). Champignons recoltes du Congo Belg. par Mme. Goossens - Fontana II, *Agaricus. Bull. Jard. Bot. Etat.* 26; 1-128. An introduction to the genus in Central Africa in preparation for the *Flore Icon.*, see pg 118. Forty-one species, many new and including the new species *Psilosace congolensis* are described, supported by line-drawings of both habit and microscopic characters. In French.

Heinemann, P. (1956-57). *Flore Iconographia Champignons du Congo*, Fasc. 5 and 6. see page 83.

Heinemann, P. (1960). Agarici Austro-Americani I. *Agaricus* of Trinidad. *Kew Bull.* 15; 231-249. Fourteen species, the majority new, are described and illustrated with line-drawings. In addition eight species (holotypes) are illustrated in colour.

Heinemann, P. (1962). Agaricini Austro-Americani II-IV. *Bull. Jard. Bot. Etat.* 32; 1- 21; 23-28 and 155-161. II. Eight species from Bolivia are described. III. Three species from Jamaica. IV. Four species from Venezuela. Several taxa are described as new and illustrated with line-drawings.

Heinemann, P. (1962). Agaricini Austro-Americani V. Etude des types C. Spegazzini. *Bull. Inst. Agron Statns. Rech. Gembloux* 30; 273-282. Sixteen taxa originally described by Spegazzini are reconsidered; only sporograms are given.

Bohus, G. (1961-79). *Agaricus (Psalliota)* Studies. I-IV. *Annls. hist.-nat. Mus. natn. Hung.* 53; 187-195; 61; 151-157: 63; 78-82: 66; 17-85: 67; 37-40: 68; 45-49. Basically in stages up-dating of Pilat's classic work, with additional information. Line- drawings supplied. For Part VII see *Sydowia Beih.* 8; 63-70.

Essette, H. (1964). Les Psalliota. *Atlas Mycol.* 1; 136; Paris. An inexhaustive account of french species of *Agaricus*; some illustrations are in colour, and are more useful than the b/w sketches provided. Descriptions of fifty-one species are given although some extralimital species are included in addition. In French.

Heinemann, P. (1965). Les Psalliotes. *Naturalistes belges*, Suppl. Bruxelles. A useful descriptive key with introductory and bibliographic information, and although intended for use in W. Europe incorporates the wide experience of the author who has studied European and extra-limital species for many years. In French.

Heinemann, P. (1965). Notes sur les Psalliotes (*Agaricus*) du Maroc. *Bull. Soc. Mycol. Fr.* 81; 372-401. Eighteen species are considered, mostly previously being described from Europe; one new combination is made and one new species described. One species previously known from S. America and one from C. America are recorded for the first time from North Africa. In French.

Heinemann, P. (1968). Some *Agaricus* from the coniferous area of the Himalayas. *Tech Comm. Nat. Bot. Lucknow* (1968); 11-21. A useful although non-illustrated contribution in modern terms from this little known area. Six species considered, half previously described from Central Africa. In French.

Issaacs. B.F. (1967). Studies on the genus *Agaricus* 1. Ag. *cretacellus* and its relationships. *Mich. Bot.* 6; 3-12. *Agaricus cretacellus* is redescribed and supported by line-drawings and b/w photographs. Its relationships are discussed.

Sheperd, C.J. (1969). Observations on Australian Agaricales I. A preliminary account of the genus *Agaricus* L. ex Fr.. *Tech. Pap. Div. Plant Ind. C. S. I. R. O. Aust.*, 27. A paper leaning heavily on European descriptions, one new combination is made. No illustrations are provided.

Heinemann, P. (1971). Quelques Psalliotes du Congo-Brazzaville. *Cah. Maboke*, 9; 5-10. Seven species are recognized from the area, three taxa were already known from C. Africa, two species are described as new and two described but remain unnamed. Line-drawings support the text. In French with English summary.

Kuhner, R. (1974). Agaricales de la zone Alpine. Genre *Agaricus* L. ex Fries. *Trav. Sc. Parc. Nat. Vanoise* 5; 131-147. Six well-known species and one new variety are described. No line-drawings are provided. In French.

Heinemann, P. (1977). Essai d'une cle de determination des genres *Agaricus* and *Micropsalliota*. *Sydowia* 30; 6-37. A descriptive key to all known species of these two genera; a comprehensive account, the culmination of many years of work by this authority on les Psalliotes. In French.

Freeman, A.E.H. (1979). *Agaricus* in North America: Type studies. *Mycotaxon* 8; 1-49. Seventy-three type specimens of *Agaricus* from North America, mainly those described by W.A. Murrill and C.H. Peck are described and nomenclatural problems discussed. Eleven further species are considered nomina dubia.

Freeman, A.E.H. (1979). *Agaricus* in southeastern United States. *Mycotaxon* 8; 50-118. A key and descriptions to forty-two species of the genus are presented; five new species are proposed.

AGARICACEAE: LEPIOTEAE = LEPIOTACEAE.
See under *LEPIOTACEAE* below.

AMANITACEAE (AMANITEAE)

GENERAL:

Heinemann, P. (1964). Les Amanitees. *Naturalistes belg., Suppl.* Bruxelles. All species of *Limacella* and *Amanita* recorded for Belgium are keyed-out. Useful b/w photographs and coloured illustrations support the key; in addition brief nomenclatorial and taxonomic notes are provided. Replaces earlier editions in 1935 and 1949. In French.

Amanita

Characters. 1. Construction of ring if present and whether friable or membranous. 2. Construction of volva if present or its replacement. 3. Shape of stipe-base. 4. Scales on pileus and stipe. 5. Striation of pileus-margin. 6. Colour changes of flesh if any.

Gilbert, J.E. (1918). *Le genre Amanita Persoon (Amanita* s. st. inc. *Amanitopsis* and *Limacella*). Etude morphologique des especes et varieties. Revision critique de la systematique. 188 pp. Lons-le-Saunier. A standard work on the genus of importance to all workers wishing to study the genus s. lato both in temperate areas and tropical localities. In French.

Beeli, M. (1935). *Flora Icon. Champignons Congo* Fasc. 1. see pg. 83.

Gilbert, J.E. (1940-1941). Amanitaceae. *Suppl. Iconographia Mycologia* (Bresadola). Descriptive data in first two parts of volume 27 and seventy-three coloured plates in third part. Many exotic species are depicted. The arrangement follows closely that used by Bresadola. Sporograms are also given. See pg. 99. In French.

Gilbert, J.E. (1924-1930). Also published many supplementary notes under the heading 'Notules sur les Amanites'. These appeared periodically in *Bull. Soc. Mycol. Fr.* although No. 30 was published separately by the author as 'Amanites d'Europe'. In French.

Vesely, R. (1933). Revisio critica Amanitarum europaerum. *Annls. Mycol.* 31; 209-304. This paper includes seventeen plates and seven figures. Much of the data forms the basis for the author's later publication see below.

Vesely, R. (1934). *Amanita. Atlas Champignons de l'Europe* I. 80 pp., Prague. A less well-known part of the Atlas; Pilat's parts are more often seen. The present publication appeared in fascicles with forty b/w plates supporting the descriptive data. In French.

Parrot, A.G. (1960). *Amanites du Sud-ouest de la France.* Biarritz, 168 pp.. Based on Gilbert's classification of seven small genera. Some b/w photographs of limited quality and one plate of sporograms support the text, although microscopic characters are generally ignored. Useful bibliography is given. In French from the Centre d'etude et de recherches scientifiques, Biarritz.

Bas, C. and Corner, E.J.H. (1962). The genus *Amanita* in Singapore and Malaya. *Persoonia* 2; 241-304. Twenty-two species are described as new; many are illustrated by coloured plates. In addition three less well-known are re-described and two obscure taxa transferred to *Amanita*. One Japanese species is described from Malaysia. Line-drawings of basidiomes and microscopic characters support the very full descriptions. A most valuable contribution.

Bas, C. (1969). Morphology and subdivision of *Amanita* and a monograph of its section *Lepidella*. Persoonia 5; 285-579. A historical survey of the delimitation of *Amanita* and its infrageneric classification is given. Morphological and anatomical characters particularly those pertaining to the volva are discussed. Keys to subgenera and sections of *Amanita* are given and a provisional monograph of *Lepidella* presented. Ninety-three species, sixteen new, are described along with three new names and seven provisional species-names. Four new combinations are offered along with keys to stirps, species, etc. within *Lepidella*. Line-drawings of basidiomes and microscopic structures support the very full descriptive data. A necessary publication for any-one working on *Amanita*.

Jenkins, D. (1977). A taxonomic and nomenclatorial monograph of the genus *Amanita* sect. *Amanita* for N. America. *Bibl. Mycol.* 57; 1-106. J. Cramer, Vaduz. Eleven species are described, two of which have several varieties. Type studies on nineteen N. American taxa are included. Text supported by twelve coloured photographs of the basidiomes and several b/w photographs of anatomical details.

Jenkins, D. (1978). A study of *Amanita* types. I. Taxa described by C.H. Peck. *Mycotaxon* 7; 23-44. Very useful descriptive data resulting from the re-examination of thirty one taxa. No illustrations.

Jenkins, D. (1979). A study of *Amanita* types III. Taxa described by W.A. Murrill. *Mycotaxon* 10; 175-200. Forty-three taxa are dealt with and microscopic data offered. No illustrations.

Limacella

Smith, H.V. (1944). The genus *Limacella* in North America. *Pap. Mich. Acad. Sci.* 30; 125-147. Twelve species are described and keyed-out including one new variety; four new combinations are made. No line-drawings; one b/w photograph of *L. glioderma* is provided. See also Kuhner, R. (1936). Recherches sur le genre *Lepiota. Bull. Sec. Mycol. Fr.* 52; 177-238. *Limacella* is included in this study.

AMANITACEAE (PLUTEEAE) = PLUTEACEAE
see pg. 40

BOLBITIACEAE

GENERAL:

Watling, R. (1965). Observations on the Bolbitiaceae. 2. Conspectus of the family. *Notes R. Bot. Gdn. Edinb.* 26 (1965); 289-323. A review of the family and its relationships to other agaric genera. A base line for later publications.

Watling, R. (1964-   ). Observations on the Bolbitiaceae: Series. with line-drawings and b/w photographs.

1. New species of *Conocybe. Notes. R. Bot. Gdn. Edinb.* 25 (1964); 309-312.
2. See above.
3. Blueing in *Conocybe, Psilocybe* and a *Stropharia* species and the detections of psilocybin. *Lloydia* 30 (1967); 151-157.
4. A new genus of Gasteromycetoid fungi. *Mich. Bot.* 7 (1968); 19-24.
5. Developmental studies on *Conocybe* with particular reference to the

annulate species. *Persoonia* 6 (1971); 281-289; Incorrectly numbered 4.

6. *Pholiota septentrionalis* and its allies. *Persoonia* 6; 313-329. 1971.
7. Validation of *Conocybe appendiculata*. Persoonia 6; 329-334. 1971.
8. Some extra-European annulate species of *Conocybe*. *Persoonia* 6; 334-339. 1971.
9. A new species of *Conocybe* with ornamented basidio-spores. Ceska Mykol. 26; 201-209. 1972.
10. The enigma of the perispore. *Notes R. Bot. Gdn., Edinb.* 34; 131-134.
11. A species of *Bolbitius* with ornamented spores. *Notes R. Bot. Gdn. Edinb.* 34; 241-243.
12. The affinities of two anomalous species. *Notes R. Bot. Gdn. Edinb.* 34; 245-251.
13. The taxonomic position of those species of *Conocybe* with ornamented basidio-spores *Rev. Mycol.* 40; 31-37. 1976.
14. *Conocybe*. sect. *Gigantae. Notes. R. Bot. Gdn. Edinb.* 35; 281-295. 1977.
15. Volvate species of *Concoybe. Sydowia, Beih* 8, 401-415.
16. On the Status of two Greenland species of *Conocybe. Astarte* 10; 57-59. 1977.
17. See below.
18. See below.

Watling, R. (1972). Bolbitiaceae; with Kits van Waveren. *Coolia* 15; 101-110. Kay to European species. In Dutch.

Watling, R. Studies in Fruit-body Development in the bolbitiaceae and the implications of such work. *Nova Hedw., Beih.* 319-346.

Watling, R. (1974). *Flore Illust. des Champignons Afr. Cent.* See pg. 84.

Watling, R. (1977). An analysis of the taxonomic characters used in defining the species of the Bolbitiaceae. Herbette Symposium on Species Concept in Hymenomycetes, 1976. *Bibl. Myc.* 61; 11-53.

Watling, R. (1980). *British Fungus Flora*: 3, Bolbitiaceae. Her Maj. Stst. Office, Edinburgh. New combinations and species described in Observations 19 and 20. *Notes R. Bot. Gdn. Edinb.* 38 (1980).

Agrocybe

Characters. 1. Presence or absence of veil (annulus or velar particles) and when present construction. 2. Taste and smell. 3. Habitat preferences. 4. Pileus-surface.

Singer, R. (1977). Key to *Agrocybe, Sydowia* 30; 194-201. Key to those world species accepted by the author is given. Several new combinations are made although some of these have already been published; one new species is described and one species ad. int. proposed.

Bolbitius

Characters. 1. Habitat preferences. 2. Colour of pileus. 3. Colour of stipe and if white.

Singer, R. (1977). Key to *Bolbitius. Sydowia* 30; 216-219. Key to those world

10

species accepted by the author. One new species is proposed and descriptive data on one other taxon included.

Conocybe

Characters. 1. Striation of pileus. 2. Colour changes of stipe with age, and degree and distribution of pubescence*. Presence or absence of veil (annulus or velar particles) and when present construction. *It is always easiest to examine the stipe in its entirety under low-power of a microscope when fresh in order to ascertain the general shape of the caulocystidia.

Kuhner, R. (1935). Le Genre *Galera*. *Encyl. Mycol.* 7; 240 pp., Paris. Illustrated descriptions of European and North African taxa. Although over forty years old this is still a standard work. Part of the text covers species now placed in *Galerina*, q.v.. Varieties and forms which appeared for the first time in this publication are now considered autonomous species.

van Waveren, Kits E. (1970). The genus *Conocybe* subg. *Pholiotina* I. European annulate species. *Persoonia* 6; 119-165. Key, full illustrated descriptions and nomenclaturial discussion on all known European annulate species of *Conocybe* are given.

Watling, R. (1964). See above under General.

BOLETACEAE (INCLUDING STROBILOMYCETACEAE)

Boletus

when taken in the widest sense. Characters. 1. Colour of sporeprint. 2. Colour and shape of tube-mouthes and whether different to the tubes. 3. Stipe-ornamentation and if present its colour. 4. Viscidness of pileus. 5. Colour changes of flesh.

GENERAL:

Covers entries which include all boletes (or boleti) either under the single genus *Boletus* (i.e. *Boletus* s. lato) or as segregate genera brought together in a single or series of texts.

Peek, D.H. (1889). Various articles on boletes of N. America in *Bull N. Y. St. Mus.* See pg.

Murrill, W.A. (1910). Boletaceae. *North American Flora*, 9; 133-161. Superseded by several later publications.

Kellenbach, F. (1926-1938). *Die Rohrlinge* (Boletaceae). *Die Pilze Mitteleuropas* 1, 138 pp, Heilbrunn. A fundamental work now superseded by Singer's updating; see below.

Gilbert, E. J. (1931). Les Boletes Livres. *Mycol* 3; 254 pp. Pans. A standard work; one of the first works introducing and defining segregate genera of *Boletus* now widely adopted by agaricologists. Certain aspects are expanded in: Notules sur les Bolets I. *Bull. Soc. Mycol. Fr.* 52; 249-250: II. Ibid. 56 (1935) 120-127. In French.

Sartory, A. and Maire, L. (1931). *Les Bolets. Monographie de genre Boletus*, Paris. A monographic treatment in two parts totalling 512 pages with 2 plates.

Snell, W.H. and Dick, E. (1932-1965). *Mycologia*: Series-Notes on Boletes I - XV (I-V and VII by Snell, also XII by Dick alone, XI with R. Singer additional, VIII with Hesler additional). *Mycologia* 24, 25, 26, 28, 33, 37,

43, 48, 50-53 and 57. Dealing with over 100 taxa of which about half are new combinations or new species. Some supporting line-drawings, b/w photographs and occasionally tabular comparisons.

Coker, W. C. and Beers, A.H. (1943). *The Boletaceae of N. Carolina*, Chapel Hill, 91 pp. An account of the boletes of S.E. United States excluding the Florida- Penninsula; illustrated by sixty-five plates, five coloured, five line-drawings of basidiospores and the remainder b/w photographs of basidiomes. Full descriptions are given. A standard work still useful to U.S. workers.

Singer, R. (1945-47). Boletineae of Florida with notes on extra-limited species. I. Strobilomycetaceae. *Farlowia* 2 (1) (1945); 97-141; with plate of line-drawings of basidia and spores. II. Boletaceae (Gyroporoideae, Sulloideae and Xerocomoideae). Ibid 2(2) (1945); 223-303. III Boletaceae (Boletoideae). *Am. Midl. Nat.* 37 (1947) 1-135 with one plate of line-drawing and one b/w photograph. IV. Lamellate Boletineae, *Farlowia* 2 (4) (1946); 527-567 with one plate of line-drawings.

A set of papers with wider implications in the taxonomy of boletes than indicated by title, including the *Clitopilus* group of agarics as the Jugasporaceae. In addition not confined by any means to the boletes of Florida or even N. America but covers discussion on many boletes from Australasia etc., most of which had never been examined before utilizing modern interpretations. Valuable for the full descriptions and keys to sections, species. etc., which support the discussional data. Many new species, infra-generic taxa and combinations are introduced for the first time. A most important contribution and essential to all working in the group luckily recently reprinted by Cramer, Vaduz. More detailed treatment than found in *Agaricales in Mod. Taxonomy* for which the articles were the basis of the part dealing with boletes. As *Bibl. Mycol.* 58. (1977). See W.A. Murrill in *Lloydia* 11 (1948); 21-35.

Pearson, A.A. (1946). Notes on the boleti, with a short monograph and key. *Naturalist* (1946); 85-99. Descriptions of all British species are given in tabular form with useful check-list epithets formerly used in Britain, one coloured plate supports text. Re-issued as British Boleti, 1946 and reprinted as part of 'Pearson's keys' with introduction and commentary by Watling, 1978.

Heinemann, P. (1951). Champignons resoltes au Congo Belge par Mme Goossens-Fontana I. Boletineae. *Bull. Jard. Bot. Etat.* 21; 223-346. An introduction to the groups in preparation for the *Flore Icon.* See below.

Heinemann, P. (1954). *Flore Iconographia Champignons du Congo*, Fasc. 3. Full descriptions supported by coloured plates of all species described for the former Belgian Congo. In French. see page 83. Fasc. 15 (1966) covers additional species.

Heinemann, P. (1954-64). Series: Notes sur les Boletineae Africaines. I. *Phlebopus* and *Xerocomus; Bull Jard. Not. Etat.* 24; 113-120. II. Trois bolets d l'Uganda: Ibid 30; 21-24: III. Ibid 34; 269-272. In French.

Singer, R. (1957). Les boletaceas austroamericanas. *Lilloa* 28; 247-268.

Pomerleau, R. (1959). Notes on Cours. *Bull. Cerele Mycol. Amateurs Quebec* 4. Suppl. Boletacees; 113-135. Keys and synopsis of East Canadian boletes. In French.

Singer, R. (1960). Les boletaceas de sudamerica tropical. *Lilloa* 30; 117-127.

Skirgiello, A. (1960). Grzyby: Fungi. Boletales. *Flora Polska Rosling*

Zarodnikowe Polski i ziem osciennych, Warsaw. In Polish. Coloured plates supporting description of Polish species.

Blum, J. (1962). *Les Bolets. Etudes Mycol.* I, Paris, pp. 164. Key to and descriptions of the majority of European boletes are given. Sixteen coloured plates and thirty-four line-drawings of basidiomes support the text., including sporograms of eight species. The coloured illustrations are from the blocks prepared for Maublanc, A. and Viennot-Bourgin, G. see pg. 75.

Kuhner, R. (1962). Notes descriptives sur le Agaricales de France. II Boletacees.. *Bull. Soc. Linn. Lyon* 31; 270-279. One species of *Paxillus* and one of *Gomphidius* are fully described along with one bolete in *'Suillus'* and one of *'Rubinoboletus'*. In French.

Singer, R. (1964). Boletes and related groups in S. America. *Nova Hedw.* 7; 93-132. Twelve species in the Paxillaceae, and four in the Boletaceae and two in the Strobilomycetaceae are described. Notes on additional species are given and keys where appropriate. Supported by two coloured plates, five b/w photographs and one plate of line-drawings. One new variety is described.

Heinemann, P. (1964). Boletineae du Katanga. *Bull. Jard. Bot. Etat.* 34; 425-478. Keys, line drawings and full descriptions supplied, many new species; See *Flore Icon.* See pg. 83.

Blum, J. (1964). Complement a trois Monographies; I. *Lactarius* et Bolets. *Bull. soc. Mycol. Fr.* 80; 281-317. Practical keys for determination. In French.

Thiers, H. (1965-77). California Boletes I-VI. (V. with Roy Halling; VI. by R. Halling alone) *Mycologia* 57-69. A series covering forty-six boletes, sixteen new, (and two varieties). Keys are supplied to *Boletus satanus* group (I), Californian Suillus spp. (III), and Californian *Leccinum* spp. (IV). see Thiers, H. (1975). *California Mushrooms* pg. 14.

Pilat, A. (1965-71). Various articles covering single species of bolete, usually supported by b/w and sometimes coloured illustrations of basidiomes *Ceska Mykol.* In Czech.

Singer, R. (1965). *Die Pilze Mitteleuropas* V, *Die Rohrlinge* I and (1967) *Die Rohrlinge* II. This is an up-dating of Kallenbach's publication produced in 1926-1938. Full colour plates depicting all species accepted by the author as occuring in Europe are supplied. Full descriptions and discussion in German although additional keys added in English. Full synonymy etc. given. Essential for bolete studies.

Blum, J. (1968-69). Revision des boletes I-IX *Bull. Soc. Mycol. Fr.* and *Rev. Mycol.* A series covering all known french taxa. I. -*B. edulis/satanus* and *purpureus* groups: II.-*B. luridius, queletii* and *erythropus* groups and *Gyroporus*: III.-*B. subtomentosus, piperatus* and *eramesinus* groups: IV. -*B. impolitus, fragrans* and *badius* groups: V.-*exannulate Suillus* spp. VI.-annulate *Suillus* spp. VII.-*Leccinum:* VIII.--*B. vitellinus, calopus* and *appendiculatus* groups.

Descriptive keys, line-drawings of habit, pileipellis and basidiospores are provided. In French.

Diek, E.A. and Snell, W.H. (1968). Some boletes from Alaska. *Mycologia* 60; 1204-1210. Twelve species, subspecies or varieties of boletes, one each in *Suillus, Xerocomus* and *Boletus*, and nine in *Leccinum* are described. One species of the third genus and four of the last genus are described as new.

McNabb, R.F.R. (1968). Boletaceae of New Zealand. N. Z. J. Bot. 6; 137-176. Twenty-two species are described and illustrated; ten species are described as new and two new combinations are provided. Keys to genera and species are given and species excludendae and nomina nuda discussed. Five plates of line-drawings and three coloured plates covering five of the new species are provided.

Smith A.H. and Thiers, H.D. (1968). Notes on Boletes. I. *Mycologia* 60; 943-954. The *B. subglabripes* and *Porphyrellus-Tylopilus* complexes are considered.

Le Clair, A. and Essette, H. (1969). *Les Boletes Atlas Mycol.*, Paris, 81 pp.. Covers the more widespread European species. Some coloured plates provided but descriptive data not exhaustive. In French.

Watling, R. (1969). *British Fungus Flora*: Part I, Boletaceae, Gomphidiaceae and Paxillaceae, Edinburgh. Full descriptions of all species recorded for the British Isles. Includes line-drawings of essential microscopic details, and keys to species. A series of publications designed for the dedicated amateur, student, etc. Based on a series of earlier papers published in *Trans. Bot. Soc. Edin.;* 39-40 (1960-67). Incorporates taxonomie and nomenclatorial information discussed in *Notes R. B. Gdn. Edinb.* 28-29. (1968-69).

Smith, A. H. and Thiers, H.D. (1971). *The Boletes of Michigan*. Ann Arbor, 428 pp.. Scientific and field keys are given to edible and poisonous boletes of the Great Lake Basin; illustrated with over three hundred b/w photographs of basidiomes and line-drawings of microscopic characters. Nine pages of additional information on type species are given. The gastroid genus *Gastroboletus* is included. Many new species are described particularly in the genera *Boletus* and *Leccinum;* several new combinations are made.

This is a scientific analysis of the boletes of a well defined area of N. America and is complementary to Singer's Boletoideae of Florida. Far more reaching than the area covered by the title.

Marchand, A. (1975-76). Quelques especes du genre *Boletus. Docums. Mycol.* 5 (21; 39-40; Ibid 6 (24); 35-40. In French. Useful additiin to Le Clair and Essette.

Snell, W. H. and Dick, E. A. (1970). *North Eastern N. America*, 115 pp. J. Cramer, Lehre. One hundred and thirty species are considered some species with additional infra-specific taxa, especially *B. edulis* (six subspecies). Full descriptions are given but the only microscopic characters offered are basidiospore sizes etc. Seventy-one coloured plates covering several species support the text; sixteen additional plates giving line-drawings of cystidia and spores are provided.

Grand, L. F. (1970). *J. Elisha Mitchell Scient. Soc.* 86; Series-Notes on North Carolina Boletes. I-III: I.*Boletellus, Phylloporus, Strobilomyces, Tylopilus,* and *Xanthoconium,* 49-56. II. *Gyrodon, Gryoporus, Xerocomus* and *Leccinum,* 57-61. III. *Suillus,* 209-213 b/w photographs support short text.

Avizohar-Hershenson, Z. and Binyamini, N. (1972-73). *Boletaceae of Israel.* Series. I. *Boletus* sect. *Luridi Trans. Brit. Mycol. Soc.* 59; 25-30. II *Gyroporus, Xerocomus* and *Boletus.* Ibid 60; 99-105. Seventeen species, one new, are so far described. Four b/w plates are provided.

Corner, E. J. H. (1972). *Boletus in Malaysia and Singapore.*pp. 263,

Singapore. (see also *Boletus* and *Phylloporus* in Malaysia; further notes and descriptions in *Gdn's. Bull. Singapore* (1974) 27; 1-16). A full discussion and introduction covering development, evolution and relationships within and between genera. Essential to all workers on boletes in Australasia. Includes line-drawings, b/w and colour photographs, sixteen coloured illustrations, keys and full descriptions to species. Many new taxa are described. *Boletus* is taken in a very wide sense.

Hongo, T. (1973). Mycological reports from New Guinea and the Solomon Islands: Enumeration of the Hygrophoraceae, Boletaceae and Strobilomyceteae. *Bull. Nat. Sc. Mus., Tokyo* 16; 543-557. Four new species and one variety described. Full description and line-drawings provided. In English.

Pilat, A. and Dermek, A. (1974). Hribovite Huby, pp. 207, Bratislava. Lavishly produced publication in Czech. Supported by innumerable colour photographs and reproductions of water-colour illustrations. Full descriptions are given. Some of the plates re-appear in the more popular publication by the same authors: Poznavajme huby, 1974 Veda.

Hongo, T. and Nagasawa, E. (1975-77). *Rept. Tottori Mycol. Inst.* (Japan): Series-Notes on some boleti from Tottori. I-IV. Fourteen species so far described and illustrated with line-drawings; one new species and one new combination are made. In English.

Heinemann, P. (1975). Les Boletinees. *Naturalistes belg.,* Suppl. 34 pp., Bruxelles. An introductory account to European boletes and allies. Short descriptive kays with illustrative material and good bibliography and specific name check-list. The genus *Boletus* is taken in its widest sense. Later edition of publication first appearing in 1961. *Nat. Belg.* 42; 332-362.

Thiers, H. D. (1975). *California Mushrooms. A field guide to the boletes*, pp. 261, New York. Keys and full descriptions of all boletes recorded for California. Supported by coloured microfiche but no line-drawings of microscopic characters. Includes twelve new species and one new variety (six species and 1 var. in *Boletus*, three species in *Leccinum*, and a single species each in *Suillus* and *Tyopilus*). See also Baroni, T. and Largent, D. (1976). New taxa of boletes of N. W. California. *Mycologia* 68; 655-661.

Grund, D. W. and Harrison, A. K. (1976). *Nova Scotian Boletus*, J. Cramer, Vaduz. Seventy-seven species and four additional varities are described for Nova Scotia. Line-drawings of basidiospores and cystidia are provided and many species illustrated in sixty-eight of the b/w photographs. Two new varieties are described. *Bibl. Mycol.* 47; 1-283.

Thiers, H.D. (1976). Boletes of the South Western United States. *Mycotaxon* 3; 261-273. A total of fifty-five species of boletes are recorded for S. W. United States; five new species are described but without supporting line-drawings.

Singer, R. and Kuthan, J. (1976). Notes on boletes. *Ceska Mykol.* 30; 143-155. One new species of *Xerocomus* and one subspecies of *Boletus* are described. The *B. erythropus* and *B. satanoides* are discussed. In English. Summary in Czech.

Horak, E. (1977). New and rare boletes from Chile. *Bol. Soc. Arg. Bot.* 18; 97-109. Three new species are described from *Nothofagus* forests of Chile and two other species are described. Five figures are provided.

Homola, R.L. and Mistretta, P.A. (1977). Ectomycorrhizae of Maine. *Life Sc.*

and *Agr. Exp. Stat., Univ. of Main Bull.* 732; 1-21. A listing of Boletaceae with their associated hosts; supported by colour photographs.

BOLETACEAE - SPECIFIC GENERA

Boletus

Vassilkov, B.P. (1966). "The White Boletus edulis", Moscow, 130 pp. An account of *Boletus edulis* including phytogeography, edibility, etc. (first twenty pages). A key to *B. edulis* and its forms, seventeen forms many new, excluding the type form, are accepted. A coloured plate covering six forms supports and descriptive data. In Russian.

Blum, J. (1968). Russules and Boletes au Salon des champignons de 1967. *Rev. Mycol.* 33; 114-136. Deals with the *Boletus satanus* and *B. purpureus* groups. In French.

Singer, R. (1977). Keys to *Boletus, Sydowia* 30; 227-253. A descriptive key to all those species of *Boletus* accepted by the author. No illustrative data included although thirteen species are discussed at length; one new species is described, and one stat. nov. and two combinations proposed.

Chalciporus formerly placed by Singer in *Suillus*.

Singer, R. (1979). Key to species of *Chalciporus. Sydowia* (1978) 31; 196-197. Key to six completely and incompletely known taxa related to *Boletus piperatus*.

Fistulinella

Guzman, G. (1974). El genero *Fistulinella* Henn. (=*Ixechinus* Heim) y las relaciones floristicas entre Mexico y Africa. *Bol. Soc. Mex. Mic.* 8; 53-63. One new species is described and illustrated. The genus is fully revised and its mycogeography considered, but see Singer's comments, and *Mucilopilus* 1979 below. In Spanish.

Leccinum

Vassilkov, B. P. (1956). Conspectus systematicus Krombholziae aurantiacae (Roques) Bilb. et hius formarum in USSR inventarum. *Mat. Otd. Spor. Rast.* 11; 134-146. Seven taxa one new are reduced to forms of *Leccinum autantiacum*(as*Krombhlzia*); supported by a single coloured plate of five forms. In Russian.

Vassilkov, B. P. (1956). *Krombholzia seabra* (Fr.) Karst. in USSR. *Trudy bot. Inst. Akad. Nauk. SSSR.* Ser. 2, *Plant Crypto.* 10; 367-384. A key to and short descriptions of twelve forms of *L. Seabrum* are given (as *Krombholzia*) with lengthy discussion on the genus in general. Five forms and the typical form of *L. seabra* now considered autonomous are illustrated in a single coloured plate. Little microscopic data presented. In Russian.

Smith, A.H., Thiers, H.D. and Watling, R. (1966). A preliminary account of the N. American species of *Leccinum* Sect. *Leccinum. Mich. bot.* 5; (3A); 131-178. ditto sect. *Leccinum*, Sects. *Luteoscabra* and *Scabra*. Ibid 6 (3A); 107-543.

A critical look at the genus *Leccinum*; description of all known N. American species, many of which are new; supported by b/w photographs, but no line-drawings of microscopic characters. Application further than the titles imply; essential to anyone studying boletes.

16

Blum, J. (1967). Essai de determination de quelques Bolets du groupe Seaber, *Rev. Mycol.* 32; 135-161 and 336-367. An account of the European species of *Leccinum* proparte, adopting a narrow species concept; care must be taken in accepting the nomenclature. Keys, line-drawings of pilei-pellis structures and shortened descriptions are given. In French.

Smith, A.H., Thiers, H.D. and Watling, R. (1968). Notes on species of *Leccinum* 1. Additions to sect. *Leccinum. Lloydia* 31; 252-267. A continuation of the study in *Mich. Bot.:* see above. Nine new species and one new variety are described with notes on *L. testaceoscabrum*. Four b/w plates of photographs are provided.

Engel, H., Dermek, A., and Watling. R. (1978). *Rauhstielrohlinge. Die Gattung Leccinum in Europe*, 76 pp. Coburg. Full descriptions, supported by coloured illustrations, line-drawings and b/w photographs. A curate's egg! In German.

Mucilopilus
Wolfe, C.B. (1979). *Mucilopilus*, a new genus of the Boletaceae with emphasis on N. American taxa.

Suillus
Smith, A.H. and Thiers, H.D. (1964). *A contribution toward a monograph of N. American species of Suillus*, pp. 116, *Ann Arbor*. Introduced by a summary of the characters and relationships of a much expanded genus (includes *Boletinus* and *Paragyrodon*). Covers forty-nine taxa including twelve new species many of which are illustrated by b/w photographs. Several excluded species are discussed including one new *Phylloporus*. A standard work with useful keys to species, sections, etc.

Blum, J. (1965). Essai de determination des bletes du groupe *Granulatus*. Bull. Soc. Mycol. Fr. 81; 450-491. An account of the exanulate European species adopting a narrow species concept; care must be taken in accepting the nomenclature. Keys, sporograms and shortened descriptions are given. In French.

Blum, J. (1965). Au salon de champignons, 1964. *Rev. Mycol.* 30; 100-111. Keys to *B. flavus-visseidus* group. In French.

Thiers, H.D. (1979). The genus *Suillus* in the western United States. *Mycotaxon* 9; 285-296. A key is presented to forty species; short descriptions of each species are given.

(Boletinus)
Singer, R. (1977). Keys to species of *Boletinus. Sydowia* 30; 227. Short key to species of the genus.

(Fuscoboletinus)
Pomerleau, R. and Smith, A.H. (1962). *Fuscoboletinus*, a new group of the Boletales. *Brittonia* 14; 156-172. A key and full descriptions are provided for six species transferred to the new genus.

(CULTURAL CHARACTERS)
Laut, G.J. (1966). Cultural characteristics of three species of *Boletinus. Con. J. Bot.* 44; 395-402. Subject as title suggests.

Pantidou, M.E. and Walton Groves, J. (1966). Cultural studies of Boletaceae: some species of *Suillus* and *Fuscoboletinus. Can. J. bot.* 44; 1371-1392.

Eighteen species of *Suillus* and three species of *Fuscoboletinus* are supplied with descriptions of their macroscopic and microscopic cultural characters. A key is supplied.

Pantidou, M.E. and Watling R. (1970). A contribution to the study of the Boletaceae; Suilloideae. *Notes R. Bot. Gdn. Edinb.* 30; 207-237. Eleven species, five new, are described in full. Cultural characters are given for all but one species. Four un-named series of collections grouped on cultural characters are described as are two of uncertain position. The relationships of all the species described with their conifer hosts and the reasons for their distribution are discussed.

Xerocomus

Thiers, H.D. (1974). The genus *Xerocomus* Quel. in Northern California. *Madrono* 17; 237-249. Keys to and full descriptions of five species recorded for this west coastal state of North America. Two plates of pileipellis-structure and sporograms are supplied.

GASTROID BOLETES:

Thiers, H.D. and Trappe, J.M. 1959. Studies in the genus *Gastroboletus*. *Brittonia* 21; 244-254. Key to ten species of *Gastroboletus* is provided with full descriptions of seven. Four species are described as new; the text is supported by two plates b/w photographs of basidiospores.

'STROBILOMYCETACEAE'

Heim, R. and Perreau-Bertrand, J. (1963). Le Genre Boletellus a Madagascar et en Nouvelle - caledonie. *Rev. Mycol.* 28; 191-199). Four species, two new, are described. One figure of line-drawings of basidiomes and one coloured plate of three taxa are supplied. In French.

Thiers, H.D. (1963). The bolete flora of the Gulf coastal plain I. The Strobilomyceteae. *J. Elisha Mitchell Sci. Soc.* 79; 32-41. Strobilomyceteae. Keys to *Strobilomyces* (2 species) and *Boletellus* (6 species) are given. Descriptions are supplied and supported by two plates of basidiospores of eight samples.

Heim, R. and Perreau, J. (1964). Deux *Boletellus* nouveaux d'Afrique tropicale. *Cah. Maboke* 2; 13-19. Full descriptions of two new species are given accompanied by two figures of line-drawings. In French.

McNabb, R.F.R. (1967). The Strobilomycetaceae of New Zealand. *N. Z. J. Bot.* 5; 532-547. Seven species and one variety are described, keyed-out and illustrated; four species are new. One coloured plate depicting four new taxa, one b/w photograph and one of line-drawings are provided.

Singer, R. (1970). *Flora Neotropica* 5; 1-32. A key to genra within the family and to species within each genus are given. Thirteen species are described; one new combination is made. The data are supported by four plates of line-drawings.

Horak, E. (1976). *Boletellus* and *Porphyrellus* in Papua-New Guinea. *Kew Bull.* 31; 645-652. Two new species and three others are fully described for the first time from this island. Four figures of line-drawings are given.

Singer, R. (1977). Key to the species of *Boletellus*. *Sydowia* 30; 221-225. A key to species known to the author is given; some additional taxonomie data is supplied.

Wolfe, C.B. (1979). *Mucilopilus*, a new genus of the Boletaceae, with emphasis

on N. American taxa. *Mycotaxon* 10; 116-131. A key, several new combinations and a new variety are proposed. Descriptive data, supported by
S.E. Micrographs and line-drawings, are given.

COPRINACEAE

GENERAL:

Romagnesi, H. (1976). Quelques especes rares ou nouvelles de Macromycetes.
Bull. Soc. Mycol. Fr. 92; 189-206, see below under *Coprinus* and *Psathyrella*. In French.

Copelandia

Weeks, A.R., Singer, R. and Hearn, W.L. (1979). A new psilocybian species of
Copelandia. J. Nat. Products 42; 469-474. *Copelandia chlorocystis* is described as new and a key to the known species in the genus given.

Coprinus

Characters. 1. Presence or absence of veil or ring, or whether min-
utely pubescent s.l. 2. Veil-characters especially s.m. (must be carried out immediately on collecting). 3. Shape and presence of pleuro-
and cheilocystidia, initially view with low power s.m. 4. Habitat
preferences.

Massee, G. (1898). A revision of the genus *Coprinus. Ann. Bot. (London)* 10;
123-184. Now completely out of date but useful as a source of a vast array
of epithets now long forgotten.

Josserand, M. (1933). Importance de l'ornamentation pileique pour la
determination des *Coprinus. Annls. Soc. Linn. Lyon.* 77; 89-95. A very
important paper in which the use of the veil-structure is discussed and
used to divide the genus. In French.

Josserand, M. (1933). L'etude des Coprins fimicoles. *Annls. Soc. Linn. Lyon* 77;
96-113. A study covering several small coprophilous members of the
genus found widespread in Europe; illustrated descriptions are supplied.
In French.

Kuhner, R. and Josserand, M. (1934). Description de quelques especes du
group de *Coprinus plicatilis* (Curt.) Fr. *Bull. Soc. Mycol. Fr.* 50; 53-63.
The first account showing the heterogeneity of the Japanese parasol
mushroom; line-drawings are given. In French.

Romagnesi, H. (1941). Les Coprins. *Rev. Mycol.* 6; 108-119. Seven species are
provided with illustrated descriptions; four new taxa are described and a
key to the C. *domesticus* group provided. In French.

Kuhner, R. and Josserand, M. (1944). Etude de quatre Coprins de groupe
'lagopus'. *Bull. Mycol. Fr. Soc.* 60; 19-32. The first account separating out
the several elements which have been lumped under the names 'lagopus'
and 'einereus' an important contribution to which geneticists should be
aware. Line-drawings are provided. In French.

Romagnesi, H. (1945). Etude de quelques Coprins (2nd). *Rev. Mycol.* 10; 73-89.
Eight species are considered including three new taxa; illustrated descriptions are given and a key to the Micacei provided. In French.

Locquin, M. (1947). Etude sur le genre *Coprinus*. Quelques coprins jimicoles.
Bull. Soc. Mycol. Fr. 63; 75-88. Seven taxa are provided with illustrated
descriptions. In French.

Romagnesi, H. (1951). Etude de quelques Coprinus 93rd). *Rev. Mycol.* 16; 108-128. A continuation of the series commenced in 1941. Six taxa all but one new are provided with illustrated descriptions. In French.

Lange, J. (1952). Species concept in the genus *Coprinus. Dansk. Bot. Ark.* 14; 1-164. A standard and most important work covering all aspects, mycological and cultural, of members of the section *Setulosi*. Keys, descriptions and line-drawings of microscopoic characters are provided.

Lange, M. and Smith, A.H. (1953). The *Coprinus ephemerus* group. *Mycologia* 45; 747-780. Twenty-three taxa of the section *Setulosi* are fully described; new species are provided with latin diagnoses but reference should be made to the article by Lange above for illustrations, etc.

Orton, P.D. (1954). Notes on British Agarics. Observations on the genus *Coprinus. Trans. Brit. Mycol. Soc.* 46; 263-276. Members of section *Lanaculi* are discussed and a key offered to the British species. Four additional species, three in section *Impexi* and one in section *Setulosi* are also provided with illustrated descriptions.

Locquin, M. (1955). Recherches sur les Coprins II. *Bull. Soc. Mycol. Fr.* 71; 5-18. Follows on from the 1947 publication; illustrated descriptions are provided for eight new species. In French.

Orton, P.D. (1961-72). Notes on British Agarics in Notes *R. Bot. Gdn. Edinb.* See page 73.

Pilat, A. and Svrcek, M. (1967). Revision specierium sectionis Herbicolae generis Coprinus. *Ceska Mykol.* 21; 136-145. Seven species, four new are d described in the *Friesii* group. A key supports the text and one b/w photograph is provided; line-drawings illustrate the descriptions. In Latin.

van Waveren, E. Kits (1968). 'The stercorarius group' of the genus *Corprinus. Persoonia* 5; 131-176. Full descriptions of all those members of the group found in Europe are given. Exhaustive illustrated descriptions and discussion are given.

Bogart, F. van de (1976-1979). The genus *Coprinus* in western N. America. I. Section *Coprinus. Mycotaxon* 4; 233-275. II. Section *Lanstuli*. Ibid 8; 243-291. III. Ibid. 10; 155-174. Section *Atramentari*. Full descriptions are provided but poor line-drawings support the text.

Romagnesi, H. (1976). Quelques especes rares ou nouvelles de Macromycetes I. Coprinacees. *Bull. Soc. Mycol. Fr.* 92; 198-200. A key to the Micacei is given and illustrated; descriptions of four new taxa are provided.

Panaeolus

Hora, F.B. (1957). The genus *Panaeolus* in Britain. *Naturalist* (1957); 77-88. A key and descriptions in tabular form are given for all British species. One plate of line-drawings is given. Reprinted as part of 'Pearsons keys' with introduction and commentary by R. Watling (1977).

O'lah, G. (1961). Le genre *Panaeolus. Rev. Mycol., Mem. Hors.* series 1. Intended to be the standard work on the genus and although full of data the author adopts an un-necessarily narrow species concept. In French.

Psathyrella   (included here may be found in addition *Lacrymaria* and *Psathyra* and even *Panaeolina*, usually merged with *Panaeolus*.)

20

Characters. 1. General stature and whether basidiomes are aggregated and/or possess a rooting stipe-base.
2. Habitat preferences whether sphagnicolous, carbonicolous, arenicolous, etc. 3. Presence of a veil specially in immature basidiomes; persistence of same or whether ring is developed. 4. Colour or gills when immature. 5. Pileus-shape and whether pileus is hygrophanous or expallant, or not.

Romagnesi, H. (1944). Classification du genre *Drosophila*. *Bull. Soc. Linn. Lyon.* 13; 15-54. see also Ibid 21; 151-156.

Romagnesi, H. (1960). Descriptions de deux nouvelles especes du sous-genre *Psathyrella*. *Bull. Soc. Mycol. Fr.* 82; 539-545.
Full illustrated descriptions are provided.

van Waveren, Kits E. (1971-1977). Notes on the genus *Psathyrella*. I-Vi. *Persoonia* 6; 249-280 and 295-312: Ibid 7; 23-54: Ibid 8; 345-406: Ibid 9; 199-231 and 281-304. A series of papers consisting of full descriptions of European species with line-drawings. Good discussional data are given. So far the following groups have been covered: *P. gracilis* complex, *P. atomata* complex, descriptions and keys to sections *Psathyrella, Ammophilae, Bipellis* and *Subatomatae;* several new taxa have been proposed and five controversial species discussed at length.

Smith, A.H. (1972). The North American species of *Psathyrella*. *Mem. N. Y. Bot. Gdn.* 24; 1-633. A most important work covering more than 410 N. American taxa includes studies in purple-spored agarics in *Mycologia* 31; 544-557 and 40; 669-682. Many new species are described; text supported by many b/w photographs and line-drawings. A mammoth treatment inclusive of *Lacrymaria* and *Panaeolina*.

Galland, M.C. and Romagnesi, H. (1975). Descriptions de quelques especes de *Drosophila* Quel. (*Psathyrella* ss. dilat.) *Bull. Soc. Mycol. Fr.* 91; 137-224. Twenty-three taxa, several new, are provided with illustrated descriptions; five un-named collections are described in addition. In French.

Romagnesi, H. (1976). Quelques especes rares ou nouvelles de Macromycetes I. Coprinacees. *Bull. Soc. Mycol. Fr.* 92; 189-198. Four species, two new, are fully illustrated and described. In French.

CORTINARIACEAE

A large family with many genera, the central one *(Cortinarius)* consisting of scores of taxa. The family is usually divided into two subfamilies separated on the colour of the spore-print., i.e. Cortinarieae and Inocybeae. For convenience the constituent genera are listed alphabetically. *Pholiota* is also included although there is evidence that although the members of the genus have brown spores they are in fact closely related to *Stropharia*.

GENERAL
See introduction to *Le Genre Inocybe* which discusses brown-spored agaries in general. see Heim. R. pg. 27.

Binyamini, N. (1973). *Inocybe* and *Cortinarius* in Israel. *Israel J. Bot.* 22; 144-150. Four species of *Inocybe* and six species of *Cortinarius* under oak and pine are described from upper Galilee and the Golan Heights. All are new to Israel. Includes a plate of line-drawings and one b/w photograph.

Binyamini, N. (1975). Cortinariaceae in Israel. *Israel J. Bot.* 24; 1-9. Eight species of *Inocybe*, eight of *Cortinarius*, one *Hebeloma* and one *Galerina* are described. all are new to Israel. All species of the family encounted in Israel this far are mentioned and briefly discussed. Two b/w photographs are included.

Alnicola. This is better called *Naucoria* as the genus is based on *Agaricus escharoides* which is also considered the type genus of *Naucoria:* see below.

Singer, R. (1978). Key to species of *Alnicola*. *Sydowia* 30 (1977); 206-209. A key to all species accepted by the author is given.

Cortinarius including **Dermocybe**.

Characters: 1. Colour of immature and mature gills. 2. Texture and distribution of veil, and presence of second veil. 3. Stature and whether emarginate stipe-base is present (always cut basidiome in half). 4. Stickiness of both pileus and stipe, or neither; in dry weather test with a little saliva or against the lips. 5. Hygrophanity or expallance of pileus.

Kauffman, C.H. (1910). *Cortinarius. North American Flora* 10 (5); 282-348. Key and descriptions of N. American species, six new species are provided with latin diagnosis.

Kauffman, C.H. (1918). *Cortinarius.* Reprinted from *Agaricaceae of Michigan*, Publ. 26; *Biol. Surv. Mich. Geol. and Biol. Survey.* One hundred and fifty-three species are described and keyed-out. Thirteen new species and one variety are introduced. Although somewhat out of date a useful catalogue of those species of *Cortinarius* described by Peck recorded for Michigan. Some determinations are tentative; Kauffman's notes are in the University Herbarium, Michigan, Ann Arbor.

Henry, R. (1935 and 1936). Etudes de quelques Cortinaires I. *Bull. Soc. Mycol. Fr.* 51; 317-340. Six species in sg. *Inoloma* and *Dermocybe* are described: no line-drawings provided. In French. II. Ibid 52; 85-99. Four species in sg. *Inoloma* and *Telamonia*; no line-drawings provided. In French.

Henry, R. (1937). Revision de quelques Cortinaires. *Bull. Soc. Mycol. Fr.* 53; 47-71. Nine species in sg. *Inoloma, Dermocybe* and *Telamonia*, one new are considered. No line-drawings are provided. In French.

Henry, R. (1937). Description de quelques *Dermocybe* du groupe *anomalus Fr. Bull. Soc. Mycol. Fr.* 53; 143-164. Four taxa, one new, are described, line-drawings of habit given. In French.

Henry, R. (1937). Etude de trois *Inoloma* et trois *Hydrocybe* dont deux nouveaux. *Bull. Soc. Mycol. Fr.* 53; 301-308. Taxa described as above; no line-drawings. In French.

Henry, R. (1939). Les Cortinaires du groupe *cinnamomeus. Bull. Soc. Mycol. Fr.* 55; 284-302. Eleven species including one comb. nov. and three new species are discussed, eight in detail. In French.

see Smith, A.H. (1939). Studies on the genus *Cortinarius* I. *Contr. Univ. Mich. Herb.* 2; 1-42. Seventy-three species are discussed, some fully documented, including fourteen new species. Twelve b/w plates illustrate twelve taxa half of which are new.

Henry, R. (1943). Essai d'une clé dichotomique analytique provisoire destinee a faciliter l'etude des Cortinaires du groupe des *Scauri (= Bulbopodium) Rev. Mycol.* 8, Suppl. 1-56. Key as title suggests.

Henry, R. (1944). Quelques especes rares ou nouvelles. *Bull. Soc. Mycol. Fr.* 60; 64-78. Seven species in *Dermocybe* and *Inoloma* are described, four new species and one new variety.

Henry, R. (1945). Essai d'une cle'... du groupe Myxacia. *Rev. Mycol.* 10, Suppl. 9-39.

Henry, R. (1946). Les Cortinaires. *Bull. Soc. Mycol. Fr.* 62; 204-218. Ten species are considered; in French.

Henry, R. (1946) Essai d'une cle'... des Phlegmacia (*Cliduchi* et *Elastici*) *Rev. Mycol.* 11, Suppl. 4-42.

Orton, P.D. (1955). The Genus *Cortinarius* I. *Myxacium* and *Phlegmacium*. *Naturalist* (July-Sept. 1955); 1-80. General notes are offered on the genus *Cortinarius* and a key to the subgenera offered. A key, tabular descriptions, systematic arrangement and discussional information in line with A.A. Pearson's keys in the same journal are provided. 102 species are included in the key, a few as yet not recorded for Britain. Of wider value than just for British Isles although the author in 1977 considered it incomplete, and did not allow it to be reprinted, with other Naturalist keys, in a photocopy edition with introduction and commentary by Watling (1978).

Orton, P.D. (1958). The Genus *Cortinarius* II. *Inoloma* and *Dermocybe*. *Naturalist* (Suppl. 1958) 81-149. General notes on the genus and its typification are given. A key, tabular descriptions, summary of classification and discussional information in line with Part I are provided. One new subgenus and seven new species are introduced. A key to subgenera and alternative keys to two small groups are included. Covers 160 species, a few of which are not as yet recorded for Britain; of wider value than just for British Isles; reprinted entirely in 1978 with short commentary by R. Watling.

Moser, M. (1960). *Die Gattung Phlegmacium*, 440 pp., Bad Heilbrunn. 54 pages of introduction with descriptions fully supported by 32 coloured plates covering 190 taxa and four line-drawings of basidiospores; plus one b/w photograph and a plate of spore-mass colours. Sixty-three taxa are provided with latin diagnoses; 168 autonomous species are dealt with. A key to subgenus *Sericeocybe* is added. In German, although keys to stirps and species are provided in addition to French.

Kuhner, R. (1959-1961). A set of several articles entitled 'Notes descriptives sur les agarics de France I. *Cortinarius*'. a) *Cortinarius myxacium. Bull. Soc. Linn. Lyon* (1959) 28; 120-127 and 131-141. Very full discussion on five taxa, three of which are supplied with new names. In French. b) (1960). ditto. *Phlegmacium*. Ibid 29; 6-13 and 40-56. Twelve species described in full, one bearing a new name and three new species. In French. .. (1960) ditto. *Phlegmacium*. Appendice aux *Phlegmacium*. Ibid 29; 65-67 *C. crassus* described and discussed. In French. c) (1960) ditto. *Inoloma* and *Dermocybe*. Ibid 29; 211-213, 219-227, 261-266. Nine species are dealth with in full. In French. d) (1961) ditto. *Telamonia* and *Hydrocybe*. Ibid 30; 50-65 and 88-101. Eight taxa are fully described. In French. e) (1961) ditto. Retouches et considerations Generales. Ibid 29; 109-113. Five species discussed. In French.

Svreek, M. (1959). *Cortinarius (Myxacium) mucifluus* Fr. et conspectus specierum sectionist Collinti Fr. *Ceska Mykol.* 13; 168-171. A key is offered to the sections; in Czech.

Bertraux, A. (1960). *Les Cortinaires. Etudes Mycol.* II. Paris, 136 pp. A useful introduction to the genus with coloured illustrations and line-drawings, but not exhaustive. Deals with large conspicuous taxa only. In French.

Singer, R. and Moser, M. (1965). Forest Mycology and Forest Communities in South America. I. *Mycopath. Myc. Appl.* 26; 178-182. A very basic review with collaborators but includes key to *Cortinarius* spp. of the Cordillera Pelada.

Henry, R. (1967-68). Etude provisiore du genre *Hydrocybe*. Hydrocybes a pied attenue a la base. *Bull. Soc. Mycol. Fr.* 83; 989-1056 and 84; 396-421. A revision of *Hydrocybe* including keys to sections and stirps and descriptions of most taxa. Fifteen new species and two new varieties are described in the second part. In French.

Moser, M. (1967). Beitrag zur Kenntnis schwarzenden Cortinarien aus der Untergattung *Telamonia*. *Schweiz Z. Filzk.* 45; 97-101. A key to two unnatural groups of eight blackening species is given with full descriptions of two, one of each is illustrated in colour. German.

Moser, M. (1967). Neue oder Kritische *Cortinarius* Arten aus de untergattung *Telamonia* (Fr.) Loüd.. *Nova Hedw.* 14; 483-518. Eighteen species, thirteen new, are described and discussed in full. Keys are given to *Helvelloidei*, stirps *Helvelloides*, *Arenatus* and *Semivestitus* and to the *Torvus* and *Evernus* groups. Latin diagnoses are given separately prior to a set of line-drawings of basidiospores and habit-sketches and for a few examples, velar tissue and cheilocystidia. In German; details on extra-limital European species given.

Svrcek, M. (1968) *Cortinarius (Telamonia) pilatii* sp. nov. und andere Arten aus der Verwandschaft von *C. flexipes* emend Kuhner. Ceska Mykol. 22; 259-278. A key to the *C. flexipes* group (Cort. section *Leptophylii* stirps *Paleiferus*) is given including three new species. Full descriptions are given in Czech and German. Line-drawings of basidiomes and basidiospores and one b/w photograph of *C. pilatii* are supplied.

Moser, M. (1969-70). *Cortinarius* Untergattung *Leprocybe* subgen. nov. die Rauhkopfe vorstudien zur einer monographie. Z. Pilzk. 35; 213-246; Ibid 36; 37-39. Discussion of new subgenus is given. Key to *Leprocybe*, stirps *Raphanoides* is provided and the new sections Brunneotincti and Bolares are proposed. A key is provided for the latter. Twelve taxa are treated with line-drawings of basidiospores. In German.

Thiers, H.D. and Smith, A.H. (1969). Hypogeous Cortinarii. *Mycologia* 61; 526-536. Four new hypogeous species of *Cortinarius* with permanent or long persisting velar membrane are described and supported by b/w photographs.

Moser, M. (1969). *Cortinarius* Fr. Untergattung *Leprocybe* subgen. nov. die Raukopfr. Z. Pilzk. 35; 213-238. The infra-classification of the new subgenus *Leprocybe* is given with key to the five sections. Keys are given to Sect. *Orellani* and Sect. *Leprocybe* (stirps, *Ignipes, Cotoneus* and *Psittacinus*) and five and seven descriptions respectively, 2 descriptions of which are to unnamed collections. Two coloured plates and a plate of sporograms are included. In German.

Raab, H. and Peringer, M. (1970—1971). *Cortinarien* - Funde in Osterreich mit besonderer Beruck eichtiguing der Ungebung Wiens. *Schweiz Z. Pilzk.* 48; 89-96 and 124-130; Ibid 49; 23-30. A continuing series of papers fully describing *Cortinarius* spp. found in Austria especially around Vienna.

24

Bohus, G. (1970-1979). Interessantere *(Cortinarius* - Arten aus dem Karpaten-Becken (Agaricales: Cortinariaceae) Series I. *Annls. hist. nat. Mus. Natn. Hung.* 62 (1970); 137-143. Twenty-three taxa, one new, are discussed; line-drawings provided. II. 68 (1976); 51-58. Three new species are described and seven others discussed; supporting line-drawings are provided. III. 71 (1979); 65-72. Descriptions of four new taxa are given; three additional taxa are discussed and two small keys are provided. Two plates of line-drawings are provided.

Moser, M. (1972-76). Die Gattung *Dermocybe* (Fr.) Wunsche (Die Hautkopfe) *Schweiz Z. Pilzk.* 50; 153-167; Ibid 51; 129-143. Ibid 52; 97-108 and 129-142. Ibid 54; 145-150. The genus Dermocybe is discussed and reviewed. An infra-classification is given and a key to the sections offered. In German; resume in French. Twenty species, four new, are described and illustrated with line-drawings.

Bon, M. and Gaugue, G. (1973-1975). Macromycetes de Belleme (Cortinarius) I. *Docums. Mycol.* 3 (11); 31-43. Six members of sg. *Phlegmacium - Cliduchi* are considered. Line-drawings are provided. In French. II. *Docums Mycol.* 6 (21); 25-37. Six members of sg. *Phlegmacium - Scauri* are considered. Line-drawings are provided. In.French.

Moser, M. and Horak, E. (1975). *Cortinarius* Fries und nahe verwandte Gattungen in Sudamerika. *Nova Hedw., Beih.* 52; 1-628. *Cortinarius* subgenera *Paramyxacium, Cystogenes* and *Dermocybe* s. g. *Icterinula* are described as new. Keys to stirps and species are supplied for all groups. Several score species are described in 182 stirps in *Cortinarius* and fifteen stirps in *Dermocybe* (of which a high proportion are new). Includes 116 plates of line-drawings and twenty-coloured plates covering 290 *Cortinarius* and 25 species of additional genera and 186 species of *Dermocybe*. In German but additional keys are supplied in English.

Henry, R. (1976). Les Cortinaires purpurescentes. *Docums. Mycol.* 6 (25); 25-55. Descriptive key to the purple-*Phlegmacia* is given; the key includes two species each of subgenera *Hydrocybe* and *Inoloma* with similar colouring. Many new taxa are proposed.

Ammirati, J.F. and Gilliam, M.S. (1975). *Cortinarius* sect. *Dermocybe*. Further studies on *Cortinarius aureifolius*. *Nova Hedw., Beih.* 51; 39-52. Two varieties, one new in addition to the type variety are described by Ammirati. B/w photographs of basidiospores, interhyphal pigment deposits and basidiomes are offered. Comparisonis made with *C. incognitus*.

Ammirati, J.F. and Smith, A.H. (1977). Studies in the genus *Cortinarius* II. Section *Dermocybe*, new north American species. *Mycotaxon* 5; 381-397. Five new species of *Cortinarius* are described from western North America. Line-drawings of spores are given.

Lamoure, D. (1977-78). Agaricales de la zone Alpine. *Genre Cortinarius* Fr. s.g. *Telamonia* (Fr.) Loud. *Trav. Sc. Parc. Nat. Vanoise* 8; 115-146, 9; 77-101. Two parts. Detailed study of thirty-one species, eight new, of small, dark hygrophanous species of *Cortinarius* is given and supported by line-drawings of basidiome habit and basidiospores. In French.

Locquin, M. (1977) *Flore Mycologique* (Mycological Flora). Vol. III and IV, Paris, 160 pp. and 156 pp. respectively, the last composed of seventy-five plates in colour and the same repeated in b/w with pertinent characters of basidiome indicated. The two volumes cover Hebelomataceae, and Cortinariaceae (Myxacoideae); the rest of *Cortinarius* is promised in the companion volumes V and VI. Vols. I & II cover the boletes and their relatives but have not as yet appeared. Published from Ecole Pratique des Hautes Etudes, Paris.

The work contains glossary, tabular descriptions, classification details, technical information as well as descriptions in latin. Many new taxa are proposed including those at levels higher than genus.

Henry, R. (1977). Nouveau regard sur les Cortinaires. *Bull. Soc. Mycol. Fr.* 93; 313-371. A critical review of the genus including the descriptions of thirteen new species. Henry's contributions can be found in several articles in *Bull. Soc. Mycol. Fr.* referred to in Orton and Moser above. Descriptive key to the genus is given in *Bull. Soc. Mycol. Fr.* 92 (1976); 52-126.

Romagnesi, H. (1977). Quelques especes rares ou nouvelles de Macromycetes III. Trav. dedies a Viennot-Bourgin 337-344. Three new species are described.

Chevassut, G. and Henry R. (1978). Cortinaires nouveaux ou rares de la region Languedoc-Cevennes (1st note). *Docums. Mycol.* 8; (32) 1-74. A series of articles covering several *Cortinarius* species found in this region of France. In French.

Cuphocybe

Horak, E. (1973). Fungi Agaricini Nova Zelandiae V. *Nova Hedw. Beih.* 43; 193-200. Three species, one new, are described and line-drawings offered.

Descolea

Horak, E. (1971). Studies on the genus *Descolea* Sing.. *Persoonia* 6; 231-248. A most comprehensive account of the genus supported by excellent descriptions and line-drawings of all the known species; four species are described as new.

Flammulaster also included in *Phaeomarasmius* see below.

Orton, P.D. (1960). New Check-list of British Agarics and Boleti. Part III. *Trans. Brit. Mycol. Soc.* 43; 232-236. Key to European species of *Flocculina* with descriptions of two new series.

Watling, R. (1967). The genus *Flammulaster. Notes R. Bot. Gdn. Edinb.* 28; 65-72. Seventeen new combinations are made in the genus with citation of specimens; one line-drawing and the description of a recent N. American collection of *Lepiota rhombospora (= Flammulaster)* are offered.

Galerina: s. lato. See Bon, M. (1971). Etudes microscopiques. *Docums. Mycol.* 1; 1-12.

Characters. 1. Copiousness of veil, whether ring or cortina. 2. Type of gill-attachment. 3. Pubescense of stipe. 4. Colour of gills and their attachment. 5. Habitat preferences.

Atkinson, G. (1918). The genus *Galerula* in N. America. *Proc. Am. Phil. Soc.* 57; 357-374. Covers both *Conocybe* and *Galerina* elements and superseded by Kuhner below in which the results are included.

Kuhner, R. (1935). *Le Genre Galera. Encycl. Mycol.* 7; 1-238 Paris. Separation of *Conocybe* and *Galerina* elements. In French. Deals with all species then known from Europe and N. Africa. Line-drawings and full descriptions provided. Although forty five years old still a very important work.

Bas, C. (1960). Notes on Agaricales II. *Persoonia* 1; 303-314. The *Pholiota marginata* group is discussed. A provisional key to the annulate field inhabiting species of *Galerina* in Europe is presented along with full descriptions of *G. uncialis* and *G. praticola*; *G. moelleri* is described as new. Line-drawings of basidiomes, basidiospores and cheilocystidia are given.

Smith, A.H. and Singer, R. (1964). *A monograph on the genus Galerina*, Earle pp. 384. New York. A taxonomic account covering 199 species is introduced by twenty-eight pages of discussion covering historical aspects, characters used in classification, both macro- and microscopic, etc. An appendix of four species discovered too late for inclusion in the main text is added along with six excluded or doubtful species. Thirty-four figures of line-drawings give cystidial characters and at the end of fifty-four line-drawings of basidiospores and twenty plates, covering sixty-eight species, of b/w photographs are provided. A standard work, important to all studying these generally musicolous agarics. Incorporates the same authors in *Mycologia* 50 (1950); 469-489.

Wells, V.L. and Kempton, P.E. (1969). Studies on the Fleshy Fungi of Alaska III. The genus *Galerina*. *Lloydia* 32; 369-387. Twenty-two species, six new, are described from Alaska. Covers species of conifer forest and bog more than tundra habitats. One figure of cheilocystidia of new species is given, and a key to species provided.

Barkmann, J.J. (1969). The Genus *Galerina* in the Netherlands. *Coolia* 14; 49-86. Fifty species, seventeen provisional and five sub-species are keyed-out in groups and additionally in an artificial key. Five species have not yet been recorded from the Netherlands. In Dutch.

Kuhner, R. (1972). Agaricales de la zone Alpine. Genre *Galerina*. *Bull. Soc. Mycol. Fr.* 88; 41-118. Three species are fully described and discussed in subgenus *Naucoriopsis* and twelve in sub-genus *Galerina*. *Bull. Soc. Mycol. Fr.* 81 (1965); 143-257. Four new species are included. Ibid 88; 119-153. Five species, three new, in subgenus *Tubariopsis* and two species of the new genus *Phaeogalera* are fully described and discussed. No line-drawings are provided. In French.

van Waveren, Kits E. (1973) *Galerina ampullaceocystis, G. cinctula* and *G. larigina*. *Proc. K. ned. Akad. Wet.*, C 76, No. 4; 392-405. *G. camerina, G. josserandii, G. pseudocamerina* and *G. larigna*, s. Smith and Singer are fully discussed and descriptions of *G. cinctula* and *G. ampullaceocystis* given. Line-drawings of microscopic characters are supplied.

Singer, R. (1974). Notes on *Galerina*. *Bull. Soc. Linn Lyon, numero special* 43; 369-405. An article bringing Smith and Singer's publication up-to-date.

Gymnopilus

Hesler, L.R. (1969). *North American species of Gymnopilus. Mycologia Mem.* No. 3 pp. 117. Covers seventy-three species of *Gymnopilus*, twenty six of which are new; some new combinations are also made and doubtful and excluded 'Gymnopilus' listed. Keys are supported by descriptions and line-drawings of basidiospores and cheilocystidia; b/w photographs are provided. One new subgenus and one new section are introduced. A compehensive account valuable to all workers in the field, particularly as it includes many European taxa.

Romagnesi, H. (1976). Sur deux especes nouvelles de *Gymnopilus. Kew Bull.* 31; 443-447. Two new European species are described; line-drawings are supplied. In French.

Hebeloma:

Characters. 1. Whether gills weep or not in fresh conditions, or whether slightly coloured drops are visible when dry. 2. Smell. 3. Whether veil is present, and if present its distribution.

Romagnesi, H. (1965). Etudes sur le genre *Hebeloma. Bull. Mycol. Soc.* Fr. 81; 323-344, two species, three new in sect. *Indusiata* and four new species in Sect. *Denudata* are fully described. Line-drawings support the information for *H. versipelle, H. mesophaeum,* and *H. crustuliniforme* groups. In French.

Moser, M. (1970). Beitrage sur Kenntnis der Gattung *Hebeloma. Z. Pilzk.* 36; 61-75. *H. funariophilum, H. perpallidum* and *H. sacchariolens* var *tomentosum* are described as new. The first is compared with *H. anthracophilum.* Line-drawings accompany full descriptions. In German.

Bruchet, G. (1970). Contribution a l'etude du genre *Hebeloma;* partie speciale. *Bull. Soc. Linn. Lyon.* Supp. 39; 1-131. Keys to subgenera, sections etc. are provided and descriptions of thirty-six species and varieties including twelve new species given. The full descriptions are supported by line-drawings of basidiomes and microscopic structures, and latin diagnoses. In French.

Bohus, G. (1972-78). *Hebeloma* Studies I. *Annls. hist. nat. Mus. Natn. Hung.* 64; 71-78. Seven species are discussed and a key given to the *H. versipelle* group. One new species is proposed and illustrated description is given. II. *Annls. hist. nat. Mus. Natn. Hung.* 70; 99-104. Two new species are described and a key offered to the *H. crustuliniforme* groupe. Line-drawings give microscopic features to four species. In English.

Hesler, L.R. (1976). New species of *Hebeloma. Kew Bull.* 31; 471-470. Illustrated descriptions for ten new species of *Hebeloma* from S.E. United States are provided.

Inocybe; includes ***Astrosporina*** as defined by Horak.

Characters: 1. Type of stipe-base (always make section never guess). 2. Whether possessing special odour or not e.g. smelling of fruit, etc. 3. Presence of lilac flush on stipe and/or pileus - make section and examine flesh at stipe-apex. 4. Whether stipe is powdered and if so whether only at apex or throughout. 5. Presence of veil or not. 6. Texture of pileus surface.

Kauffman, C.H. (1921). Studies on the genus *Inocybe. Bull. N. Y. State Mus.* 223-224; 43-60.

Sartory, A. and Maire, L. (1923). Synopsis du genre *Inocybe.* 246 pp., Paris. A compilation of species-diagnoses in alphabetical order gleaned from various authors. Two poor plates of line-drawings which are not really helpful are provided.

Boursier, J. and Kuhner, R. (1928). Notes sur le genre *Inocybe. Bull. Soc. Mycol. Fr.* 44; 170-189. Full description of members of the *I. lanuginosa* group are given; line-drawings are supplied with a full discussion on the names of all species suspected to be in the complex.

Heim, R. (1931). *Le Genre Inocybe. Encly. Mycol.* 2, Paris, 1-429. A fundamental work not only on *Inocybe* but a review of all brown-spored genera of agarics preceding full descriptions of known *Inocybe* taxa.

28

Supported by line-drawings and coloured plates. Essential to all those studying *Cortinarius* and related taxa. In French.

Kuhner, R. and Boursier, J. (1932-33). Notes sur le genre *Inocybe*: I. Les *Inocybe* goniosporees. *Bull. Soc. Mycol. Fr.* 48; 118-161. II. Ibid 49; 81-121. Nine species, two new, are described in each part and in the second are supported by an analytical key. Line-drawings of cystidia and basidiospores are offered in each part and much useful discussional data given. In the first part a key to the section Cortinatae is offered.

Pearson, A.A. (1954). The Genus *Inocybe*. *Naturalist*, (Oct.-Dec. 1954); 117-140. Unfinished manuscript completed by R.W.G. Dennis. A useful introduction to the genus in Britain with key, descriptive tables and very helpful discussional information. Dealing with sixty-five species of the then-known British taxa. Reprinted in 1978 with commentary by R. Watling but with no alternations to text.

Stangl, J. and Veselsky, J. (1971). Beitrag zur Kenntnis der selteneren *Inocybe* Arten, *Ceska Mykol.* 25; 1-9, 27; 11-25, 28; 138-142 and 143-150. 28; 65-78 and 195-218. 30; 65-80 and 70-175. 31; 15-27 and 189-192. 32; 22-31 and 161-166. 33; 68-80, and 137-137. Continuing series. Thirty-five species, eight new are discussed up to going to press. Keys, full illustrated descriptions are given, some illustrations in colour. Contributions VIII in 30; 170-175 gives a key to twenty-two species. In Czech, with summaries in German.

Grund, D.W. and Stuntz, D.E. (1969-77). Nova Scotia Inocybes I. *Mycologia* 60; 406-425. Collections from N. Scotia include two new species and two new to N. America. II. Ibid 62; 925-939. Eight species collected are described of which one species and one variety are new; only one has been previously reported before. III. Ibid 67; 19-31. Six species, five new, and two varieties are reported for the first time. IV. Ibid 69; 392-408. Nine species are considered and two varieties described as new. B/w photographs and line-drawings support descriptive data.

Stangl, J. (1973-1975). Uber einige Risspilze Sudbayerns I. *Z. Pilzk.* 37; (1973); 17-40. II. Ibid 39 (1975); 191-202. Good descriptions with line-drawings and coloured illustrations. In German.

Stangl, J. (1976-78). Die eckigsporigen Risspilze. I *Z. Pilzk.* 40 (1976); 65-80. II. 42 (1978); 15-32. Good descriptions supported with line-drawings and good coloured illustrations. In German.

Horak, E. (1977). Fungi Agaricini Novaezelandiae VI. *Inocybe* and *Astrosporina N. Z. J. Bot.* 15; 713-747. Twenty four new taxa are described all of them forming ectotrophic mycorrhiza with either *Nothofagus* spp. (Fagaceae) or *Leptospermum* spp. (Myrtaceae). Neither genus was previously recorded from N. Zealand. A key and line-drawings for all species are provided.

Horak, E. (1979). *Astroporina* (Agaricales) in Indomalaya and Australasia. *Persoonia* 10; 157-205. Key, descriptions and line-drawings of thirty species, twenty one new, known from the India-Australia region are given. One new variety is also described. Insufficiently known species etc. are discussed in addition to the formal descriptions. The genus is equivalent to *Inocybe* subg. *Clypeus*.

Bon, M. (1979). Inocybes rares, critiques ou nouveaux dans le Nord de la France. *Sydowia, Beih.* 8; 76-97. Twelve taxa, seven new, are described in full; several new infrageneric divisions are proposed. Line-drawings support the text. In French.

Romagnesi, H. (1979). Quelques especes rares ou nouvelles de Macromycetes III. *Inocybe. Sydowia, Beih.* 8; 349. Eight species, five new (and one new form) are provided with full illustrated descriptions.

Kuehneromyces

Singer, R. and Smith, A.H. (1946). The taxonomic positon of *Pholiota mutabilis* and related species. *Mycologia* 38; 500-523. Four species, two new, are admitted into the new genus *Kuehneromyces* with *P. mutabilis* as type. The new family Strophariaceae is proposed and latin descriptions supplied for all new taxa. Full descriptions accompanied by b/w photographs are given.

Naucoria s. lato: see also *Alnicola* and *Simocybe.*

Characters. 1. Smell and taste. 2. Whether the pileus is striate or not. 3. Whether the pileus-colours are in red-browns or yellow-browns. 4. Whether the stipe is tough or possesses a rooting base. 5. Presence or absence of a veil. 6. Overall size.

Orton, P.D. (1960) New Check List of British Agarics and Boleti. Part III. *Trans. Brit. Mycol. Soc.* 43; 308-327. A key to British brown-spored agarics placed in the traditional genus *Naucoria* is given. The key includes British species of *Simocybe* as well as *Naucoria* s. st. (= *Alnicola*). Ten new species are provided with full illustrated descriptions.

Phaeocollybia:

Smith, A.H. (1957). A contribution towards a monograph of *Phaeocollybia. Brittonia.* 9; 195-217. Sixteen species, seven new, are described and illustrated with line-drawings. Keys are provided to members of the two sections accepted. (see *Mycologia* 64; 1138-1153, where six further species are described with J.M. Trappe.).

Singer, R. (1970). *Phaeocollybia* (Cortinariaceae). *Flora Neotropica* 4; 1-11. Four species, one new, are described. Line-drawings are provided along with key to sections, some new, and to species within these sections.

Horak, E. (1977). Further additions towards a monograph of *Phaeocollybia. Sydowia* 29; 28-70. An account of all known taxa referrable to *Phaeocollybia* based on fresh collections and examination of herbarium material. A key is provided with line-drawings. Eleven new species are described (see also *Acta Bot. Ind.* 2 (1974); (69-73) amongst thirty-four species dealt with; one new combination is proposed. Incorporates information available in Fungi Agaricini Novae Zelandiae IV. see pg. 83.

Phaeogalera

Kuhner, R. (1972). Agaricales de la Zone Alpine. *Bull. Soc. Mycol Fr.* 88; 143-151. Two species are fully described for this newly proposed genus; no line-drawings are provided. In French.

Phaeomarasmius

Singer, R. (1956). Versuch einer Zusammenstellung Arten der Gattung *Phaeomarasmius. Schweiz A. Pilzk.* 34; 44-65. Covers species not only in

Phaeomarasmius but taxa considered by some to be placed in *Flammulaster* (*Flocculina* Kuhn. and Romagn.) and some even in *Pholiota* (see Hesler and Smith below).

Watling, R. (1967). see *Flammulaster*.

Pholiota

Characters: 1. Veil whether forming a cortina or ring. 2. Odour if present. 3. Habitat preferences. 4. Gelatinous or scaly pileus, or combination of both.

Smith, A.H. and Hesler, L.R. (1968). *The North American species of Pholiota.* pp. 402, Ann Arbor. One hundred and fifteen b/w illustrations of basidiomes and forty line-drawings of basidiospores and cystidial structures support the descriptions; two hundred and five species are considered, many new. Five doubtful species mostly belonging to *Phaeomarasmius* are also dealt with. A full list of excluded species and their current taxonomic position is given in an appendix as are three species formerly ascribed to *Flammula* and described from the West Indies by Patouillard. The genus is accepted in a very broad sense and contains *Kuehneromyces*, *Pachylepyrium* and *Flammulaster* as well as the genus *Pholiota* as outlined by Singer. A standard work with much discussional data; keys are supplied to sections, stirps and species and the twenty page introduction covers the characters used in identification.

Pleuroflammula

Horak, E. (1978). *Pleuroflammula. Persoonia* 9; 439-451. Ten species of the genus are described and illustrated by line-drawings. Taxonomic and geographic--distributional data is given.

Rozites

Moser, M. and Horak, E. (1975). *Cortinarius* Fries und nahe Verwandte Gattungen in Sudamerika. *Nova Hedw., Beih.* 52; 513-519. Summary of the genus with key and descriptions of two new species is given. In German; Key in English (pg. 607). Key supplied supersedes Moser Die Gattung *Rozites* in *Schweiz Z. Pilzk.* 31; 164-171.

Stephanopus

Moser M. and Horak E. (1975). *Cortinarius* Fries und nahe Verwandte Gattungen in Sudamerika. *Nova Hedw., Beih.* 52; 520-523. Descriptions of the new genus and two species are given. In German; key in English (pg. 608).

CREPIDOTACEAE

Crepidotus

Pilat, A. (1948). Monographie des especes europeanes du genre *Crepidotus* Fr. Atlas des Champignons de l'Europe 6, Prague, 84 pp. Twenty-four species are accepted, one of uncertain position. Four species would now be placed in *Melanotus, Paxillus, Pleuroflammula* and *Simocybe*. Twenty-four b/w photographs are presented. One volume in a series of valuable works. In French.

Singer, R. (1947). Contributions toward a monograph of the genus *Crepidotus*. *Lilloa* 13; 59-95. A key to species is offered and many classic species discussed. Several new sections are described. As the title suggests this is a contribution not a monograph, but nevertheless an important one.

Pilat, A. (1950). Revision of the types of the extra-european species of the genus *Crepidotus*. *Trans. Brit. Mycol. Soc.* 33; 215-249. A key is given to eighty-three species including seven new species. Discussion and illustrations are given for sixty taxa, many little known except for the type collection.

Orton, P.D. (1960). New Check List of British Agarics and Boletes III. *Trans. Brit. Mycol. Soc.* 43; 218-221. A key to British species is offered with brief notes on a few entries; one new species is described as new.

Singer, R. and Moser, M. (1965). Forest mycology and forest communities in South America I. *Mycopath. Mycol. Appl* 26; 183-184. A very basic review prepared with collaborators but includes key to species of *Crepidotus* sect. *Defibulati* in S. America.

Hesler, L.R. and Smith, A.H. (1965). *North American species of Crepidotus* 168 pp., New York. A standard work. One hundred and twenty-five species are descirbed, many new. Fourteen N. American species are excluded and placed in *Melanotus, Pleuroflammula, Phyllotopsis, Phaeomarasmius, Pyrrhoglossum* and *Simocybe*. More far reaching than title suggests. B/w photographs and line-drawings of basidiospores and cystidial structures accompany the descriptions. See addition as by first author in *Nova Hedw. Beih.* 51; (1975); 133-137. The photographs are repeated at the back because of the poor quality within the text.

Singer, R. (1973). Monograph of Neotropical species of *Crepidotus*. *Nova Hedw., Beih.* 44; 345-484. A key is offered to sixty-two tropical species; seventeen new species and fifteen new varieties are described. Line-drawings are provided.

Hesler, L.R. (1975). New species of *Crepidotus. Nova Hedw., Beih.* 51; 133-137. Four new species are described; no illustrations are supplied.

Horak, E. (1977). *Crepidotus episphaeria* and related species from the southern hemisphere. *Ber. Schweiz Bot. Ges.* 87; 227-235. Five species of *Crepidotus* with thick-walled and crystal bearing cheilocystodoa are described from S. Africa, Java, New Caledonia and N. Zealand.

Melanomphalina

Singer, R. (1970). A revision of the genus *Melanomphalina* as a basis of the phylogeny of the Crepidotaceae. In *Evolution of the Higher Basidiomycetes*, Knoxville. Seventeen species, thirteen new, are described and a key to them offered; a full discussion of the relationships of *Melanomphalina* is given.

**Simocybe*

Romagnesi, H. (1962). Des *Naucoria* des Groupe *centunculus (Ramicola)*. *Bull. Soc. Mycol. Fr.* 78; 337-358. A key to seven European species is given all being transferred to *Agrocybe*. One new species is proposed; line-drawings are provided; in French.

Singer, R. (1973). Neotropical species of *Simocybe, Nova Hedw. Beih.* 44; 485-517. Eighteen species, three new, are described and a key offered. Line-drawings are presented.

Tubaria

Romagnesi, H. (1940). Essai sur le genre *Tubaria. Rev. Mycol.* 5; 29-43. A very useful contribution to an understanding of the members of this small genus; line-drawings supporting. Six species are described and keyed-out and one new variety proposed. In French. Useful additions in *Rev. Mycol.* 8; 26-35. with an up-dated key.

ENTOLOMATACEAE

The family is often defined as a single genus (*Rhodophyllus* or *Entoloma*) by some authors. For convenience this is adopted.

> Characters. 1. General stature, including presence or absence of stipe. 2. Sealiness and degree of striation of pileus. 3. Attachment of gills. 4. Presence or absence of blue-colours.
> (Also s.m. shape of basidiospores, pigmentation of hyphae and presence or absence of clamp-connections).

Romagnesi, H. (1941). Les Rhodophylles de Madagascar. *Prodrome Fl. Mycol. Madag.* No. 2, Paris

Romagnesi, H. (1956). Champignons recoltes du Congo Belge par Mme Goossens. Fontana. *Rhodophyllus. Bull. Jard. Bot. Etat.* 26; 137-182. Description of species in preparation for the *Flore Icon,* see below.

Romagnesi, H. (1957). *Flore Icon. Champignons Congo, Rhodophyllus.* Fasc. 6; 131-137. see pg.

Hesler, L.R. (1963). A study of *Rhodophyllus* types. *Brittonia* 15 (4); 324-366.

Hesler, L.R. (1967). *Entoloma (Rhodophyllus)* in S.E. North America. *Nova Hedw., Beih.* 23; 1-192. A standard work covering many N. American taxa; not restricted to S.E. United States but important to workers throughout N. America and other temperate countries. See name corrections in *Mycologia* 64 (1974); 717 and Morgan Jones in Can. J. Bot. 49 (1971); 1052.

Largent, D. (1970-74). Studies in the Rhodophylloid fungi. I. Generic concepts *Madrono* 21; 32-39. A full discussion of the genera within the *Entolomataceae* is presented (with R.G. Benedict). II *Albolephonia,* a new genus. *Mycologia* 62; 437-452. Eleven taxa in the newly introduced genus *Alboleptonia* are described (with R.G. Benedict.) III.
IV. *Leptonia* sect. *Leptonidii. N.W. Sci.* 48; 52-56. *Entoloma jubatum* group is discussed. V. *Leptonia* subgenus *Paludocybe* section *Albicaules* and sect. *Rosecaules* and related taxa. *N. W. Sci.* 48; 57-65. Two new taxa are described and four new combinations made. Fifteen species are discussed and a key to members of section *Albicaules* given.

Kuhner, R. (1971). Agaricales de la zone Alpine; Rhodophylaceae. *Bull. Soc. Mycol. Fr.* 87; 9-13: Ibid 87; 433-444. An introduction to the group with descriptive data based on personal collections. Full descriptions are given only; no line-drawings are provided. In French.

Largent, D. (1971-   ). Rhodopohylloid fungi of the Pacific Coast (United States). I. Type studies and new combinations of species described prior to 1968. *Brittonia* 23; 238-245. Fifteen taxa are described, of which six are transferred to other genera within the Entolmataceae. II. New or interesting subgeneric taxa of *Nolanea*. *N. W. Sci.* 46; 32-39 (with H. Thiers). Four new sections are proposed for the genus *Nolanea* and three new species and two new varieties in *Nolanea* are presented; descriptions are provided. IV. Infrageneric concepts in *Entoloma*, *Nolanea* and *Leptonia*. *Mycologia* 66; 987-1021. Keys to genera of rhodophylloid fungi as well as infrageneric taxa of *Entoloma*, *Leptonia* and *Nolanea* are offered with discussion. One new species and several new sections are described; two new combinations are also made.

Horak, E. (1973). Fungi Agaricini Nova Zelandiae I. *Nova Hedw. Beih.* 43. Forty-nine species and one each in *Richoniella*, *Claudopus* and *Pouzaromyces*, the last new, are fully described, line-drawings support the text.

Largent, D. (1974). New or interesting species of *Claudopus* and *Entoloma* from the Pacific Coast. *Madrono* 22; 363-373. *Claudopus byssideus* and several taxa within the *E. madidum* and *E. trachyosporum* (new species) groups are described.

Romagnesi, H. (1974). Etude de quelques Rhodophylles. *Bull Soc. Linn Lyon* 43, Spec. No. 365-387. Nine species, six new, are provided with fully illustrated descriptions. In French.

Horak, E. (1975) . On cuboid-spored species of *Entoloma* (Agaricales). *Sydowia* 28; (1-6): 171-237. Also see Additions to "on cuboid-spored species of *Entoloma*" Ibid 29; (1-6), 289-299. A major contribution to the knowledge of all 'rhodophylls' with cuboid spore-shape. *Entoloma* is used in the wide sense. Forty-three, fourteen new and seven taxa, three new are described respectively. Keys are given in the first account.

Mazzer, S.J. (1976). A monographic study of the genus *Pouzarella*. a new genus in the Rhodophyllaceae. *Bibl. Mycol.* 46 192 pp. A wide ranging account of thirty taxa originally in *Nolanea* and *Leptonia*. The new genus *Pouzarella* of debatable validity is introduced for *Pouzaromyces*. Includes line-drawings etc.

Romagnesi, H. (1976). Quelques especes. . . I. Les *Rhodophyllus* sect. *Lampropodes*. *Bull. Soc. Mycol. Fr.* 92; 291-303 (also see *Bull. Soc. Linn. Lyon* 43; 325-352). Five species are dealt with; one species and one variety are described as new in the *L. lampropus* complex. In French.

Largent, D. (1977). The Genus *Leptonia* on the Pacific Coast of the United States. *Bilb. Mycol.* 55, 286 pp. Full descriptions are given to 136 members of the benus *Leptonia* recorded from the Pacific Coast of N. America (United States). Keys are given to sections, to species and infraspecific ranks. Supporting b/w photographs of basidiomes and some micro-graphs are supplied but line-drawings of microscopic characters are lacking. A most important manual which is important to all workers in temperate countries.

Pegler, D.N. (1977). A revision of the Entolomataceae from India and Sri Lanka. *Kew Bull.* 32; 159-220. A full revision of all the types of agarics from the area and referable to the family and housed at Kew are given.

Line-drawings support the revision and a key is provided.

Romagnesi, H. and Gilles, G. (1978). Rhodophyllees de Forets Cotieres du Bagon et de la Cote d'Ivoire. *Nova Hedw. Beih.* 59. An enormous compilation of members of the Entolomataceae from this western area of Africa but more far-reaching than the title would suggest at first sight. *Rhodophyllus* embraces all the segregate genera. In French.

Horak, E. (1978). *Entoloma* in South America. *Sydowia* 30; 40-111. A most important contribution covering all known species of the Entolomataceae, *Entoloma (= Rhodophyllus)* known from South America and environs. Full descriptions, descriptive data and line-drawings are supplied. Many new species are proposed.

Noordeloos, M.E. (1979). *Entoloma* subgenus *Pouzaromyces* emend in Europe. *Persoonia* 10; 207-243. A revision of known European taxa is given based on collections from European herbaria. Eleven taxa, three new, are recognized; they are all described, keyed-out and illustrated with line- drawings. Four new combinations are made. The subgenus is defined and *Pouzaromyces* emended. This replaced Moser's contribution in *Persoonia* 7; 281-288.

Noordeloos, M.E. (1979). Type studies on entolomatoid species in the Velenovsky herbarium - I. *Persoonia* 10; 245-265. Thirty species described by Velenovsky are re-examined resulting in eighteen new combinations and four new names. One described in *Nolanea* has been transferred to *Pluteus*. A separate index is provided.

Arnolds, E.J.M. and Noordeloos, M.E. (1979). New taxa of *Entoloma* from grasslands in Drenthe, the Netherlands. *Persoonia* 10; 283-300. Fourteen new species and three new varieties of *Entoloma* from grassland vegetation are described. Some line-drawings and taxonomic discussion are presented.

Horak, E. (1980 in press). *Entoloma* in Indomalaya and Australasia. *Nova. Hedw., Beih.* 80; 1-319. 234 taxa are considered the majority supported by illustrated descriptions. 234 figures and eight b/w plates are provided.

GOMPHIDIACEAE
Characters.   1. Texture of veil. 2. Colour of flesh and young gills.

Singer, R. (1946). The Boletineae of Florida with notes on extra-limital s cies, lamellate families (Gomphidiaceae, Paxillaceae and Jugasporaceae). *Farlowia* 2 (4); 527-537. Part of a wide-ranging review article on the boletes and their allies. Superseded by 1949 article. Key to subgenera and taxa in Florida. Two species described with the second with several subspecies.

Singer, R. (1946). The Boletineae of Florida with notes on extra-limital spe-426-489. A key to subgenera and sections is given; thirteen species are described with numerous infraspecific taxa outlined. Up dates and expands Kauffman, C.H. in *Mycologia* 17 (1925); 113-126.

Watling, R. (1969). British Fungus Flora: Boletaceae, Gomphidiaceae and Paxillaceae, Edinburgh, HMSO. A key to and full descriptions of all British species is given; six species are dealt with, one with some hesitation; one species *C. helveticus* is new to Britain but at a later date *(Notes*

Roy. Bot. Gdn., Edinburgh 30 (1970); 391-394 was described as a new
species: C. corallinus Miller and Watling. C. brittanicus was added by
Khan and Hora (Trans. Brit. Mycol. Soc. 70; 155-156).

Chroogomphus
Miller, O.K. (1964). Monograph of Chroogomphus (Gomphidiaceae).
 Mycologia 56; 526-549. Key and full descriptions to all fully known species
 are given; less completely known species are discussed. Illustrated with
 b/w photographs of basidiomes and microscopic details. Eight species are
 considered.
Singer, R. and Kuthan, J. (1976). Notes on Chroogomphus (Gomphidiaceae).
 Ceska Mykol. 30; 81-89. Discussion on C. helveticus and C. rutilus, where
 two subspecies are recognized in the first; one additional subspecies is
 recognized in C. rutilus. A revised key to all species known is given.

Gomphidius
Miller, O.K. (1964). The genus Gomphidius with a revised description of the
 Gomphidiaceae and a key to the genera. Mycologia 63; 1129-1163. Three
 sections containing nine species and two varieties are described for Gomphidius. A key, line-drawings, photo-micrographs and b/w photographs
 of basidiomes are given. Two Species, one section two varieties are described as new; the genus Cystogomphus is included.

HYGROPHORACEAE

Hygrophorus is taken in the widest sense.
 Characters. 1. Presence of veil. 2. Ornamentation of the stipe. 3.
 Vissidness of stipe and pileus if smooth or if scaly, colour of scales. 4.
 Colour and attachment of gills. 5. Compactness and wateriness of
 flesh and any colour changes.

Dennis, R.W.G. (1953) Some West Indian Collections referred to Hygrophor-
 ous. Kew Bull. (1953); 253-268.
Orton, P.D. (1960). New Check List of British Agarics and Boleti, Part III.
 Trans. Brit. Mycol. Soc. 43; 246-271. Key to British species of subgen.
 Hygrocybe and Camarophyllus is given. Fourteen species, nine new, are
 described; two species of subgen. Hygrophorus are in addition discussed.
Hesler, L. and Smith, A.H. (1963). North American species of Hygrophorus
 pp. 416, Knoxville (also see addit. in Sydowia 8; 304-333). A standard
 work with discussional information introducing full descriptions of all
 known N. American species and many new taxa.B/w photographs and
 some line-drawings accompany description. Keys to species, sections and
 stirps are provided. A very necessary work for all working on the Hygrophoraceae; Hygrophorus is taken in its widest sense. Supersedes publications: Studies in N. American species of Hygrophorus I and II (1939) by
 same authors in Lloydia 2; 1-62 and Ibid 5 (1942); 1-92.
Heinemann, P. (1963). Chapignons recoltes au Congo par Mme Goossens-
 Fontana, V. Hygrophoraceae Bull. Jard. Bot. Etat. 33; 421-458. A intro-

ductory account preparing information for *Flore Champignons*, see below. A key to genera and illustrated descriptions of thirteen species are provided. Three new species are described and many new combinations proposed.

Heinemann, P. (1966) *Flore Icon. Champignons Congo*. Fasc. 15. see pg. 83.

Arnolds, E. (1974). Note on *Hygrophorus*. I. *Persoonia* 8; 99-104. II. Ibid, 8; 239-256. Series. So far two new species are described as new. *H. marchii* and four closely related speceis are described and a key to the group offered.

Singer, R. (1977). Key to the speceis of *Camarophyllus*. *Sydowia* 30; 271-277. A key to all the species accepted by the author is presented; two new species described. Restricted as the title implies to species of *Camarophyllus; H. pratensis* grp.

Ben, M. (1976). Le Genre *Hygrocybe*. *Docums. Mycol.* 6 (25); 1-24. Descriptive key to sections and to species within the genus as outlined by Orton and Watling (*Notes R. Bot. Gdn., Edinb.* 29; 129-138) is offered.

Kuhner, R. (1976-78). Agaricales de la zone Alpine. *Bull. Soc. Mycol. Fr.* 92; 455-515; Ibid 93; 55-115 and 117-144. An extensive study important to all working in temperate countries; species not restricted to montane regions. Many descriptions and much discussional data based on field material given; several new species described. Unfortunately no line-drawings of microscopic details given. In French.

Pegler, D. and Fiard, J.P. (1978). *Hygrocybe* sect. *Firmae* in Tropical America. *Kew bull.* 32; 297-312. An illustrated revision of the *Firmae* group (characterised by dimorphic basidia) which occur in Tropical America. Nine taxa are recognized. Two new species, one new name and one new variety are described; three new combinations are made in *Hygrocybe*.

Singer, R. and Kuthan, J. (1978). Einige interessantes europaische Hygrophoraceae. *Z. Pilzk.* 42; 5-13. A useful contribution in German. Three species of *Hygrocybe*, two new, one species of *Camarophyllus* and species of *Hygrophorus* are considered. Line-drawings and b/w photograph support the text. In German.

Bird, D.J. and Grund, D.W. (1979). Nova Scotian species of *Hygrophorus*. *Proc. Nova Scotian Inst. Sci.* 29; 1-131. Fifty three species and varieties, two new, *H. lignocola* and *H. macrosporus*, are described new for Nova Scotia. B/w photographs and line-drawings of spores and basidiomes support the descriptions. One species is recorded for the first time from Nova Scotia. The classification adopted and species interpretation is strongly based on Hesler and Smith (see above). Assemblages around *H. olviaceoalbus, H. puniceus, H. coccineus, H. miniatus,* and *H. cantharellus* are discussed and a new series adopted for the *H. marginata* group. *H. lignicola* is placed in the new section *Lignicolohygrophorus*.

Kuhner, R. (1979). contribution a la connaissance du genre *Hygrocybe* (Fries) Kummer. Quelques recoltes de la zone silvatique. *Sydowia, Beih.* 8; 233-250. Six species, one new, are described and correlated with nuclear number of basidiospores.

Arnolds, E. (1979). Notes on *Hygrophorus*. III. The group of *Hygrophorus olivaceoalbus (Hygrophorus* subsect. *Olivaceoumbrini* Bat.) in north-western Europe. *Persoonia* 10; 357-382. Extensive, illustrated descriptions of the four species of *Hygrophorus* in this group are given. A key to the species supports the nomenclatorial and taxonomic data. One new species is described. Particular attention is paid to the ecology and description of the species.

LEPIOTACEAE (=AGARICACEAE: LEPIOTEAE)
(see Boiffard, J. (1973). Etudes microscopiques. . .
Docums. Mycol. 3 (8); 39-45.)

The genera within this family are often considered to constitute a large genus with members exhibiting a wide-range of basidiome-morphology. Purely for a convenience *Lepiota* is taken herein in a wide sense. *Cystoderma* in Tricholomataceae, q.v.

The following genera are accepted by many authors as independent genera: - *Chamaemyces* = *Drosella*; *Leucoagaricus*; *Leucocoprinus*; *Macrolepiota*; *Lepiota* s. stricto.

> Characters. 1. Consistency and structure of veil on pileus and stipe (and type of components composing the veil s.m.). 2. Colour changes when handled and whether flesh changes colour on exposure to the air. 3. Sealiness of pileus and stipe. 4. Colour of gills both when immature and mature.

Kauffman, C.H. (1922). Genus *Lepiota* in the United States. *Pap. Mich. Acad. Sci.* 4; 311-344. Short descriptions with keys; although over fifty years old still not superseded. Four new species are described.

Beeli, M. (1932). Fungi Goassensiansi IX. Genre *Lepiota. Bull. Soc. Bot. Belg.* 64; 2 ser. 14; 206-219. Descriptions supported by two plates. Information later incorporated into Flora Icon Champignons Congo. see pg. 83.

Kuhner, R. (1936). Recherches sur le genre *Lepiota. Bull. Soc. Mycol. Fr.* 52; 17-238. A basic work on the genus. The range of variation found in the genus is demonstrated although *Lepiota* is taken in its widest sense and even includes *Cystoderma*. A key to sections and species is given and line-drawings and sporograms support full descriptions of many taxa some of which are new species or new names. In French.

Huijsman, H.S.C. (1943). Observations sur le genre *Lepiota. Meded. ned. Mycol.Vereen.* 28; 3-58 with twelve figures. In French.

Romagnesi, H. and Locquin, M. (1944). Notes sur les Lepiotes I. *Bull. Soc. Mycol. Fr.* 60; 38-42. Species with green colours in the pileus are discussed and keyed-out, and members of the *L. setulosa* complex considered. In French.

Locquin, M. (1951). Les Lepiotes du sous-genre *Lepiotula* etudies des especes francais. *Bull. Soc. Linn. Lyon* 20; 153-157. Four species, one new, are keyed-out and described. In French.

Locquin, M. (1951). Premiere liste des especes descrites dans le genre *Lepiota* s. lato *Bull Soc. Mycol. Fr.* 67; 365-382. A useful reference to the original descriptions of all *Lepiota* spp. described up to 1951. In French.

Locquin, M. (1951-52). Les especes francaises du genre *Leucocoprinus* -

38

Part I, Sect. Procarae. *Rev. Mycol.* 16; 213-234. Ibid 17; 47-54. Keys and descriptions of species now placed in *Macrolepiota* are given; one new species described. In French.

Dennis, R.W.G. (1952). *Lepiota* and allied genera in Trinidad, British West Indies, *Kew Bull.* 9; 459-500. One new *Amanita* and several descriptions of non-lepiotoid fungi are given in addition to keys and descriptions to thirty-four species of *Lepiota* s. lato. Twelve new species are included in the descriptions; two new combinations and one state nov. are provided along with line-drawings.

Beeli, M. (1936). *Flora Icon. Champignons Congo.* Fasc. 2. *Lepiota* see pg. The first of a pair of publications on Tropical African agarics. The field notes are based in the main on data collected together by Mme. Goossens-Fontana.

Smith, H.V. (1954). A revision of the Michigan species of *Lepiota. Lloydia* 17; 307-328. Descriptions and keys to Michigan species are given; one new combination is made. Supersedes Kauffman's account in 'Agaricaceae of Michigan' although covers many widespread N. American taxa. No line drawings.

Locquin, M. (1956). Quelques Lepiotes nouvelles au critiques. *Friesia* 5; 292-296. *L. helveola* group is discussed and described; a new species is proposed. In French.

Babos, M. (1958-74). Studies on Hungarian *Lepiota* I-IV. *Annls. hist. nat. Mus. Natn. Hung.* 50; 87-98. 53; 195-199; 61; 157-164; 66; 67-75: Series. A useful analysis of several groups of *Lepiota* spp. or individual taxa.

Huijsman, H.S.C. (1961). *Lipiota* Sect. *Micaceae* J.E. Lange, *Schweiz Z. Pilzk.* 4; 1-7. Ten European species are discussed and keyed-out.

Aberdeen, J.E.C. (1962). Notes on Australian *Lepiota* in the Kew herbarium. *Kew Bull.* 16; 129-137. Twenty-three taxa including *Xerulina asprata* all collected prior to 1900 and housed at Kew are discussed.

Smith, H.V. (1966). Contribution toward a monograph on the genus *Lepiota* I. Type studies in the genus *Lepiota. Mycopath. Myc. Appl.* 29; 97-117. Forty-nine holo-type collections are described; includes studies on Morgan's material described in *J. Mycol.* 12-13, and compliments Kauffman (1924); see above.

Heinemann, P. (1967). *Flore Icon. Champignons Congo.* Fasc. 16; *Chlorophyllum.* see pg. 83.

Heinemann, P. (1968). Le genre *Chlorophyllum* Mass. (Leucocoprinaceae). Apercus systematique et description des especes congolaise. *Bull. Jard. Bot. Nat. Belg.* 38; 195-206. Full discussion of the genus is given, covers *C. molybdites* and its satellites. In French.

Heinemann, P. (1969). Le genre *Macrolepiota* Sing. (*Leucocoprinaceae*) au Congo-Kinshasa. *Bull. Jard. Bot. Nat. Belg.* 39; 201-226. A key to eleven species, including eight recorded from Congo-Kinshasa is given. Two new species and three new combinations are made; descriptions are given.

Heinemann, P. (1970). *Flore Icon. Champignons Congo* Fasc. 17: *Macrolepiota.* see pg. 83.

Pegler, D.N. (1972). A revision of the genus *Lepiota* from Ceylon. *Kew Bull.* 28; 155-202. Revised illustrated descriptions are provided for the type collections of all species of *Lepiota* described by Berkeley and Broome from

Ceylon. The species are distributed in three genera of which thirty-four are recognized in *Lepiota*, four in *Leucocoprinus* and two in *Macrolepiota*. A full synonmy and key are included.

Bon, M. and Boiffard, J. (1972). Lepiotes des dune Vendeenes. *Bull. Soc. Mycol. Fr.* 88; 15-28. Two new arenicolous species, one *Lepiota* and one *Leucocoprinus* are described; three new combinations are also made and a short key to *Leucocoprinus* group 'rubentes' given. In French.

Heinemann, P. (1973). Leucocoprinees nouvelles d'Afrique Centrale. *Bull Jard. Bot. Nat. Belg.* 43; 7-13. Nine new species, one new section, one new genus and one new name are all proposed along with seven new combinations. In French.

Bon, M. and Boiffard, J. (1974). Lepiotes de Vendee et de la Cote Atlantique Francaise I. *Bull. Soc. Mycol. Fr.* 96; 287-306. In French.

Heinemann, P. (1973) and (1977). *Flore Illust. Champignons Afrique Centrale.* Fasc. 2. Further contributions to the Flore. see pg. 84.

Heinemann, P. (1977). Leucocoprinees nouvelles d' Afrique Centrale I. *Bull. Jard. Bot. Nat. Belg.* 47; 83-86. Preparation of material for the *Flore Illust.* see above.

Bon, M. (1977). Les Lepiotes de l' Herbier Boudier. *Bull. Soc. Mycol. Fr.* 7; 11-22. A very useful analysis of species of *Lepiota* in the Boudier herbarium. In French.

Knudsen, H. (1978). Notes on *Cystolepiota* Sing. and *Lepiota* S.F. Gray. *Bot. Tidsskr.* 73 (2); 124-136. Generic limits are discussed and expanded to cover two sections formerly placed in *Lepiota* s. lato. Two new species are described, five new combinations made and one new name provided. A key is also provided to sect. *Echinatae*.

Babos, M. (1979). The species of the 'Rubentes' group in the genus *Leucocoprinus Sydowia, Beih.* 8; 33-53. Seven species of the *Lepiota badhami* complex are provided with full illustrated descriptions.

PAXILLACEAE

For Hygrophoropsis also see Tricholomataceae; for *Paxillus* see below.

Hygrophoropsis

Heinemann, P. (1963). Champignons recoltes au Congo per Mme. M. Goossens-Fontana IV. *Hygrophoropsis. Bull. Jard. Bot. Etat.* 33; 413-415. One new species is described with comments on *H. aurantiaca*. In French.

Paxillus and Phylloporus

Those with an asterisk include *Hygrophoropsis* and/or *Cheiminophyllum*. Characters. 1. Character of stipe. 2. Reaction with aqueous solution of ammonium hydroxide. 3. Thickness, colour and degree of branching of gills.

Singer, R. (1946). *The Boletineae of Florida Part IV. *Farlowia* 2; 527-567: see above under Gomphidiaceae. Part of a wide ranging review article on the boletes and their allies. Key to species and higher taxa within *Paxillus* and *Hygrophoropsis* is given; three species of the former and two of the latter with notes on extra-limital species are considered.

Singer, R. (1946). *Monographs of S. American Basidiomycetes especially those of the East slopes of the Andes and Brazil VI. Boletes and related groups in S. America. *Nova. Hedw.* 7; 93-132. Ten species in five genera are described; one new variety and two species in five genera are described; one new variety and two species of *Paxillus* of uncertain position are described.

McNabb, R.F.R. (1969). *The Paxillaceae of New Zealand. *N.Z. J. Bot.* 7; 349-363. Six species are described, four of which are new; keys and illustrated descriptions of genera and species are given.

Watling, R. (1969). *British Fungus Flora. Boletaceae, Gomphidiaceae* and *Paxillaceae.* Edinburgh, HMSO. A key to and full descriptions of all British species are given; three species in *Paxillus* and one in *Phylloporus* are described and supported by line-drawings.

Corner, E.J.H. (1970). *Phylloporus* Quel. and *Paxillus* Fr. in Malaya and Borneo. *Nova. Hedw.* 20; 793-822. Thirteen new species and two varieties of *Phylloporus* are described, along with two new species of *Paxillus.* Eight coloured plates cover most of these species; three further previously published species are described. Many other species-names are listed and some are discussed particularly in respect to their geographic distribution. Keys are offered to species for both genera in S. E. Asia.

PLUTEACEAE (=AMANITACEAE: PLUTEEAE)

Chamaeota

Singer, R. (1979). Key to species of *Chamaeota. Sydowia* (1978); 31; 197-198. A key to the genus and modern description of a Schulzer (1866) taxon transferred to the genus are given.

Pluteus

Characters. 1. Stipe whether scaly, punctate, granular or with darker fibrils. 2. Pileus whether hygrophanous and/or striate. 3. Gill-edge darker or concolorous. 4. Colour of flesh. 5. Whether terrestrial or lignicolous.

Vacek, V. (1948). The Bohemian and Moravian species of the genus *Pluteus. Studia Bot. Csl.* 9; 30-48. A useful publication now superseded by contributions of Singer; see below.

Singer, R. (1956-59). Contribution toward a monograph of the genus *Pluteus. Trans. Brit. Mycol. Soc.* 39; 145-232. Ibid 42; 223-226. The types of most species described in *Pluteus,* and in other genera but attributed to *Pluteus,* are discussed. The best known species of the genus are set-out in an indented key. Fayod's classification is adopted although modified with the introduction of several additional microscopic characters. A very important work.

Singer, R. (1958). Monograph of South American Basidiomycetes. I. *Lloydia* 21; 195-229 (with supplement in *Sydowia* 15 (1961); 112-132) sixty-six species, the majority new, are described excluding varieties some also new; numerous line-drawings and one b/w photograph accompany the text. A list of incompletely known species and excluded species and index are given. The supplement covers seventeen additional species, six new and five infra-specific categories; line-drawings are added.

Smith, A.H. and Stuntz, D. (1958). Studies on *Pluteus* I. *Lloydia* 21; 115-136. Thirty-two species described from North America are redescribed based on examination of type material. No illustrations are provided.

Orton, P.D. (1960). New Check-list of British Agarics and Boleti. Part III. *Trans. Brit. Mycol. Soc.* re; 343-367. Key to all those species recorded for the British Isles is given. Eighteen species are discussed of which six are proposed as new. Line-drawings accompany the text.

Horak, E. (1964). Fungi Austroamericani. II *Pluteus. Nova Hedw.* 8; 63-199. Thirty-two species, the majority new (ten in sect. *Trichoderma*; eight in sect. *Hispidoderma* and fourteen in Sect. *Celluloderma*), are described and accompanied by twenty plates of line-drawings giving all relevant characters. Keys to species are provided.

Homola, R.L. (1972). Section *Celluloderma* of the genus *Pluteus* in North America. *Mycologia* 64; 1211-1247. Keys to subsections and species within them are given. Twenty-four species and varities are accepted including one new species and two provisional taxa. One plate of line-drawings and six b/w photographs support the text.

Horak, E. (1978). *Flora Illust. Champignons Afric. Central.* Fasc. 5 see pg. 83. See also: Neue zairische Arten aus der Gattung *Pluteus*. in *Bull. Jard. Bot. Nat. Belg.* 47; 87-89.

Volvariella

Characters. 1. Colour of volva both outside and within. 2. Viseidness of pileus. 3. Substrate preferences.

Beeli, M. (1935). *Flore Icon. Champignons Congo.* Fase. 1. see pg.

Heim, R. (1936). Les Volvaires. *Rev. Mycol.* 1, Suppl; 55-58 and 85-90. An introductory treatment of the genus including a table for the determination of the European species. Line-drawings of hbait of basidiomes are given.

Shaffer, R. (1957). *Volvariella* in North America, *Mycologia* 49; 545-579. A key is presented to eighteen species recorded for N. America. Nine new combinations, one state nov. and one species nov. are proposed. Five further imperfectly known species are discussed. Three plates of line-drawings giving microscopic details accompany the text.

Shaffer, R. (1962). Synonyms, new combinations and new species of *Volvariella. Mycologia* 54; 563-572. Twelve species, two new, are discussed and several described in full. Six new combinations are made. Two figures of line-drawings are supplied.

Orton, P.D. (1974). The European species of *Volvariella. Bull. Soc. Linn. Lyon* Suppl. 311-326. A key and check-list are provided to European species. One new species is described. This is an updating of the same author's key in *Trans. Brit. Mycol. Soc.* 43 (1960); 283.

Heinemann, P. (1975). Observations sur le genre *Volvariella. Bull. Jard. Bot. Nat. Belg.* 45; 185-193. Morphological and cultural characters of the Padi-straw fungus are given; one new species and one new combination are made. One plate of line-drawings and one b/w photograph support the information.

Pathak, N.C. (1975). New species of *Volvariella* from Central Africa. *Bull. Jard. Bot. Nat. Belg.* 45; 195-196. Four new species are described; in

42

preparation for the *Flore Illust.* see below.
Heinemann, P. (1975). *Flore Illust. Champignons Afric. Central.* Fasc. 4 see
pg. 83.

RHODOPHYLLACEAE SEE ENTOLOMATACEAE

RUSSULACEAE

GENERAL:
Heim, R. (1942). *Les Lactario-Russules du domaine Oriental de Madascar.*
Prod. Fl. Mycol. Madagascar, Paris (1937). This publication drew attention to a whole new group of *Russula* spp., inclusion of which into the then
accepted system was difficult. Keys to and descriptions of taxa (twenty-
seven in *Russula* and thirteen in *Lactarius*), many new, are given with
discussional data on classification in general. Numerous line-drawings of
microscopic characters and basidiomes support text. Four coloured
plates illustrating twenty-three taxa are given; three b/w photographs
are also offered. In French.
Chiu, W.F. (1945), The Russulaceae of Yunnan. *Lloydia* 8; 31-59. Thirty-one
species of *Russula* and nineteen of *Lactarius*, of which three of the former
and five of the latter are considered new, are described. Microscopic
details are depicted in line-drawings.
Schaeffer, J., Neuhoff, W. and Herter, W.G. (1949). Die Russula een Best-
immungstabelle fur die mitteleuropaisehen *Russula* and *Lactarius*-
Arten. *Sydowia* 3; 150-173. Up-dating of observations published by the
first two authors in *Pilze Mitteleuropas.* See below, pg. 43 and pg. 45.

Lactarius
Characters. 1. Taste and smell. 2. Colour of milky latex when fresh
and after some time. 3. Nature of pileus-viseid, downy, shaggy or
zoning.
A good bibliography is found in Hesler and Smith (1979) see
below.

Romagnesi, H. (1949). Recherches sur le Lactaires de la section des Fuliginosi
Konrad. *Rev. Mycol.* 14; 103-114. A very useful contribution to a rather
perplexing group of *Lactarius* spp. with pale milk which reddens on
exposure to air either immediately or after some time.
Pearson, A.A. (1950). The genus *Lactarius. Naturalist* (1950); 81-91. An
introduction to the genus with 'key' spore types listed and figured in a
single line-drawing. A key is provided to all British species of *Lactarius*
with descriptions in tabular form. Useful list of comments on specific
epithets used in Britain is included. Reprinted in 1978 as The Naturalist
Agaric Keys by A.A. Pearson with introduction and commentary by R.
Watling.
Heim, R. (1950). Notes systematiques sur les Champignons du Perche II, Les
Lactarius a lait rouge. *Rev. Mycol.* 15; 65-79. One article of two with
line-drawings and coloured illustrations on the *L. deliciosus* group. Two
new species are proposed.

Tuomikoski, R. (1953). Die *Lactarius*-Arten Finlands. *Karstenia* 2; 9-32. Thirty-seven species briefly dealt with; useful ecological information but not supported by illustrations. In German.

Heim, R. (1955). Les Lactaires d'Afrique intertropicale. *Bull. Jard. Bot. Etat.* 25; 1-91. An extensive article in preparation for the *Flore Icon.* with descriptions, coloured and line illustrations; several new taxa are proposed and in all twenty-one species considered with full supporting discussional data. In French.

Heim, R. (1955). *Flore Icon. Champignons Congo.* Fasc 4. see pg. 83.

Neuhoff, W. (1956). *Die Michlinge: Pilze Mitteleuropas*, Vol IIB, Bad Heilbrunn, 248 pp. Standard work on European Lactarii; illustrated with beautiful coloured plates. Full monographic treatment not superseded. In German.

Babos, M. (1959). Notes on the occurrence in Hungary of *Lactarius* species with regard to the range in Europe. *Annls. hist. nat. Mus. Natn. Hung.* 51; 171-196. In English. Information on the occurrence and frequency of *Lactarius* spp. in Hungary is given based on herbarium data and field notes. A key is offered to all those species known from Hungary. No line-drawings but useful ecological data are provided.

Heinemann, P. (1960) Les Lactarius. *Naturalistes belg.* 41; 133-156. 2nd edition (1st edition in 1948). A most useful descriptive key with bibliographic data and species-notes. Three b/w photographs and one line-drawing are provided. In French.

Hesler, L. and Smith, A.H. (1960-62). Studies of *Lactarius*: I-III Series. I. Section *Lactarius. Brittonia* 12; 119-139. II. Sections *Scrobiculati, Crocei, Theiogali* and *Vellus*, Ibid 12; 306-350; III. Section *Plinthogali*, Ibid 14; 369-440: Full illustrated descriptions of North American species are offered. Information contained in this series is incorporated in Hesler and Smith (1979): American species of *Lactarius*. 841 pp. Ann Arbor. Illustrated with over 400 line-drawings; this book covers all the known North American taxa. Includes and b/w photographs including selection of micrographs. See below.

Blum, J. (1964). Les Lactaires du groupe *aurantiacus. Rev. Mycol.* 29; 101-114. A very useful discussion with descriptions and sporograms of species mentioned. In French.

Blum, J. (1965). Au salon du Champignon de 1964. *Rev. Mycol.* 30; 89-111. Remarks on the *L. vellereus* group. In French.

Blum, J. (1966). Lactaires et Russules au salon des Champignons de 1965. *Rev. Mycol.* 31; Suppl. 85-106. A useful key to the *L. vellereus* group and to *L. pallidus, hysginus* and *trivialis* complex. Sporograms offered; in French.

Blum, J. (1966). *Les Lactaires* du groupe *Piperatus. Bull. Soc. Mycol. Fr.* 82; 241-247. Notes and part key in French. Continuation of full key in *Bull. Soc. Mycol. Fr.* 80; 285-296.

Breisinsky, A. and Stangl, J. (1970). Beitrage zur revision M. Britzelmyr's 'Hymenomyceten' *Lactarius. Z. Pilzk.* 36; 41-59. A study as the title suggests; useful compilation in German.

McNabb, R.F.R. (1971). The Russulaceae of New Zealand 1. *Lactarius* DC. ex S.F. Gray. *N.Z.J. Bot.* 9; 46-66. Eight species and one variety occuring in

New Zealand are described and illustrated by line drawings of microscopic characters. Six taxa are described as new and a key offered. One
new section is proposed.

Romagnesi, H. (1974). Etude sur les Lactaires de la sous-section des *Striatini.
Bull. Soc. Mycol. Fr.* 90; 139-146. Very useful contribution to this small
group of often overlooked species around *L. eyathula*. In French.

Kuhner, R. (1975). Agaricales de la zone Alpine, Genre *Lactarius* S.F.
Gray. *Bull. Soc. Mycol. Fr.* 91; 5-69. A further important contribution in
this series covering the same ground in *Lactarius* as in *Russula* (see below
pg. 47.) Unfortunately no line-drawings accompany the wealth of data
presented. In French.

Blum, J. (1976). Les Lactaires. *Etudes Mycol.* No. 3. 371 pp. Paris. An account
of the more common European *Lactarius* spp. with descriptions and
sporograms. Some poor black and white but useful colour illustrations
are provided; accompanied by a key. In French.

Singer, R. (1976). Tropical Russulaceae; *Lactarius* sect. *Polyspaerophori.
Nova. Hedw.* 26; 897-901. Two species fully described and one illustrated
by coloured photographs.

Hesler, L. and Smith, A.H. (1979). *North American species of Lactarius.* 841
pp. Ann Arbor. Illustrated with over 400 line-drawings and b/w photographs. An account dealing with all known N. American species including many new taxa; in addition some European taxa thought to have been
found in N. America or expected to occur there are discussed. Full
descriptions following an informative introduction covering micro- and
macro-characters, historical review of the genus and ecology of the
members are provided in common with other publications by these
authors. In addition an appendix with b/w electron micrographs has been
prepared by R. Homola and N. Weber and therefore the book is the first
monographic treatment to incorporate large scale SEM studies. Six subgenera, eight sections and five subsections are defined in which over two
hundred and fifty species are distributed. The necessary keys are supplied. Replaces earlier contributions by G.S. Burlingame in *Mem. Terrey
Bot. Club* 14; 1-109 (with fifteen figures).

Pegler, N. and Fiard, J.P. (1979). Taxonomy and ecology of *Lactarius*
(Agaricales) in the Lesser Antilles. *Kew Bull.* 33; 601-628. Keys and
illustrated descriptions are provided for the Carribean species of *Lactarius* with special reference to those of the Les. Antilles. Six new species
are described and one colour-plate included; line-drawings support the
text.

Russula

See Bon, M. (1971). Etudes microscopiques. . . *Docums. Mycol.* 2; 1-12.
Characters.   1. Taste and smell. 2. Reaction of flesh with $Fe\,SO_4$ and
sulpho-vanillin. 3. Pileus-colour (to be recorded in detail). 4. Consis-
tency of flesh. 5. Whether pileus margin enrolled or not, or fluffy-
woolly. 6. Colour of thick spore-print - leave overnight.

Maire, R. (1910). Les bases de la classification dans le genre *Russula. Bull. Soc.
Mycol. Fr.* 26; 49-125. See H. Romagnesi, *Les Russules d' Europe* . . .
below.

Melzer, V. and Zvara, J. (1928). Ceske holubinky & Russulae Bohemiae). Flore monographique des Russules de Boheme. *Bull. Soc. Mycol. Fr.* 44; 135-146. A French summary of the Czech book noted in the title and published earlier in *Archiv. prirod vyzk Cech.* 17; 1-212. An analytical key to species is given. A classic work, including earlier publications e.g. *Bull. Soc. Mycol. Fr.* 59; 61-71.

Singer, R. (1926). Monographie der Gattung *Russula. Hedwigia* 66; 163-260. In German. See R. Singer below.

Crawshay, R. (1930). *Spore ornamentation of Russula* pp. 173 London. A classical publication on the taxonomic importance of spore-studies and the way this should be conducted. Short descriptions accompanied by sporo-grams are given for many European taxa. Care must be taken with the interpretation of species accepted by Crawshay by reference to later publications. Two colour-plates supplied.

Singer, R. (1931-1936). Contribution a l'etude des Russules I. *Bull. Soc. Mycol. Fr.* 46; 209-212. II. Ibid 52; 111-114. Useful notes continuing the author's earlier studies. In French.

Singer, R. (1932). Monographie der Gattung *Russula. Beih. Bot. Zbl.* 49. (II); 205-380. Continuation of publication in 1926. I - seventy-three species are described; no line-drawings given and only one b/w illustration offered. II - seventy species are described; index and bibliography given. See also *Z. Pilzk.* 3; 73-78 and 107-112. Ibid 5; 13-18 and 73-80. In German.

Schaeffer, J. (1933). Bestimmungstabells fur die europischen Taulblinge. *Z. Pilzk.* 12; 48-53 and 83-91. In German.

Singer, R. (1936). Les Russules de L'Herbier de Boudier. *Rev. Mycol.;* 19-24. Redeterminations of Boudier's names in tabular form, no line-drawings given. In French.

Niolle, P. (1938-1949). Various *Russula* studies. I. *Bull. Soc. Mycol. Fr.* 54 (- 3rd part) 3-7. II. *Rev. Mycol.* 4; 81-86: III. *Bull. Soc. Mycol. Fr.* 55; 303-307. IV. *Bull. Soc. Linn. Lyon* 9 (1940; 51-56. V. *Annls. Mycol.* 39 (1941); 66-76. VI. *Annls. Mycol.* 41 (1943); 299-302. VII. *Bull. Soc. Mycol. Fr.* 60 (1942); 101-106. VIII. *Bull. Soc. Mycol. Fr.* 64 (1948); 252-253. IX. *Bull. Soc. Mycol. Fr.* 65 (1949); 85-92. X. *Schweiz Z. Pilzk.* 8 (1949); 86-89. Mimeographed notes were also distributed on a personal basis after 1949.

Schaeffer, J. (1935). Les Systeme naturelle des Russules. *Bull. Soc. Mycol. Fr.* 41; 263-276. Information incorporated into 'Pilze Mitteleuropas,' see below.

Perason, A.A. (1948). The genus *Russula. Naturalist* (June-Sept. 1948). 2nd edition; reprinted separately 1950. One of a series of publications in which all British species of a genus are keyed-out and their morphological characters arranged in a tabular form. A very useful list of notes on species not in the tables or of dubious or controversial nature is given. Superseded but republished in 1978 with commentary and introduction by R. Watling.

Heinemann, P. (1950). Les Russules. *Bull. Soc. Nat. Oyonnax Suppl.* No. 4, 3rd edition, see *Naturalistes belg.*, (1944 2nd edition). A useful descriptive key to European species then known (and accepted). An equally useful compilation of species-names with comments on validity and synonymy etc. is given. In French.

Singer, R. (1950). Les Russules de L'Argentine. *Rev. Mycol.* 15; 125-137. Six species, four new, are described with a small key separating the taxa. No line-drawings are offered.

Blum, J. (1951-52). Quelques especes collectives de Russules. *Bull. Soc. Mycol. Fr.* 67; 167-172 and Ibid 68; 225-257. Remarks on *R. queletii* grp. (67) and *R. nauseosa* and *turei-amoena* grps. (68) are considered. In French.

Schaeffer, J. (1952). *Russula-Monographie. Pilze Mitteleuropas* II. Bad. Beilbrunn pp. 295. Standard work on European Russules; illustrated with beautiful coloured plates and supported by full, exhaustive descriptive data with discussional comments; see Healer 1961 below. This supersedes the same author's *Russula*-monographie in *Annls. Mycol.* 31 (1933); 305-516: Ibid 32; (1934) 83-91. Reprinted 1970 by J. Cramer.

Singer, R. (1952). Russulaceae of Tinidad and Venezuela. *Kew. Bull.* (1952); 295-302. A paper based on the collections made by R.W.G. Dennis in the Carribean and Venezuela. See Dennis: Add. *Kew Series* pg. 92.

Blum, J. (1953). Russules rares ou nouvelles. *Bull. Soc. Mycol. Fr.* 69; 430-450. Various species are considered. In French.

Blum, J. (1960-61). Russules complements. *Bull. Soc. Mycol. Fr.* 76; 244-274. Includes a study of *R. delica* group. Ibid 77; 153-182. In French.

Blum, J. (1960-61). A propos de Russules. *Rev. Mycol.* 25; 126-138 and Ibid 26; 190-209. In French.

Hesler, L.R. (1960). A study of *Russula* types I. *Mem. Torrey bot. Club* 21; 1-59. One hundred and fifty-seven taxa are considered and basidiospores, cystidial and pileipellis structures given from a reworking of the types; all are N. American taxa. Three hundred and eighteen figures of line-drawings support the text.

Hesler, L.R. (1961). A study of *Russula* types II. *Mycologia* 53; 605-625. The second paper devoted to the results from a re-working of further N. American types of *Russula* spp. Thirty-four taxa are reported and supported with line-drawings.

Hesler, L.R. (1961). A study of Julius Schaeffer's Russulas. *Lloydia* 24; 182-198. Thirty-five collections of classic European species have been selected for analysis and treated in a similar way to earlier publications; data supported by line-drawings. Because of the substantial role played by Schaeffer's publication in classifications his material has been examined.

Blum, J. (1962-63). Les Russules au Salon des Champignons I. - 1961 *Rev. Mycol.* 26; 71-74: II. - 1962 Ibid 27; (1963) 107-112 - on red capped species: III. - 1963. Ibid 27 (1963); 224-234 - discussion on *R. nigricans* groups. In French.

Blum, J. (1962). Les Russules, *Encyl. Mycol.*, Paris, 228 pp. An account of major numbers of *Russula* spp. recorded for France and neighboring countries. In French; superseded by Romagnesi, see below.

Singer, R. (1963). Four interesting European Russulae of subsection *Sardoninae* and *Urentinae* sect. *Russula. Sydowia* 16; 289-301. Useful discussional data on *R. sardonia* and *R. adulteriana* groups.

Shaffer, R. (1962). The subsection *Compactae* of *Russula. Brittonia* 14; 254-284. A key is given to the twelve members, four new, of the subsection found in N. America. One species *R. densifolia* is segregated into several forms mostly new. Full descriptions are given, line-drawings and eight b/w photographs support the text. Imperfectly known taxa are discussed.

Shaffer, R. (1964). The subsection *Lactarioideae* of *Russula*. *Mycologia* 56'; 202-231. Arranged similarly to companion publication with key, descriptions and illustrations. Ten species, three new, are considered; one new variety is proposed. Imperfectly known taxa are discussed.

Romagnesi, H. (1967). *Les Russulas d'Europe et de l'N. Afrique*. Bordas, pp. 998. A standard work profusely illustrated with line-drawings of microscopic details; one colour plate for use with spore-prints. Full descriptions, discussion, synonyms, etc. are given. A most useful work of importance outside even the large area covered in the title; northern European species characteristic of alpine areas, however, are largely ommitted. Comprehensive bibliography; many varieties etc. considered under several taxa and species ad int incorporated.

Rayner, R.W. (1968-70). Keys to British species of *Russula* 1-3. *Bull. Brit. Mycol. Soc.* Published originally as parts with additions etc. in Vol. 10 (2) 69-73; now available as a separete. One hundred species are considered in the original publication with twelve further species in the additions. Part I. Keys to the British species of *Russula:* Part II. Identification of cystidial structures and basidiospores: Part II. Lists of species; specific diagnostic features presented.

Heim, R. (1970). Particularities remarquables des Russules tropicales *Pellisculariae lillputiennes* - les complexes *annulata* and *radians*. *Bull. Soc. Mycol. Fr.* 86; 59-77. (See also *Boissiera* 7 (1943); 266-280). Five previously described (*Rev. Mycol.* 34; (1969)) taxa are considered and supported by one coloured plate (photographs and paintings), b/w photographs and line-drawings.

Shaffer, R. (1970). Notes on the subsection *Crassotunicatinae* and other species of *Russula*. *Lloydia* 33; 49-96. Taxonomic treatment of this exlusively N. American subsection of *Ingratae* with descriptions and keys. Descriptions and discussion on fourteen additional species from various sections of *Russula* concludes the work. Very useful line-drawings are supplied.

Shaffer, R. (1970). Cuticular terminology in *Russula*. *Brittonia* 27; 230-239. Apart from the topic of the title *R. crutosa* and *R. virescens* are described and illustrated by line-drawings.

McNabb, R.F.R. (1973). Russulaceae of New Zealand. 2. *Russula* Pers. ex. S.F. Gray. *N. Z. J. Bot.* 11; 673-730. Thirty-two species of *Russula* occuring in New Zealand associated with *Nothofagus* and *Leptospermum* are described and illustrated by line-drawings of microscopic characters. Two coloured plates depicting four species are given, along with a key.

Shaffer, R. (1972). North American Russulas of the subsection *Foetentinae*. *Mycologia* 64; 1008-1053. A key and description of ten almost exclusively N. American species within the subsection are given including one new species. Line-drawings of pertinent microscopic characters are given. A very welcome contribution to an understanding of this difficult yellow brown capped group.

Kuhner, R. (1975). Agaricales de la zone Alpine. Genre *Russula* Pers. ex S.F. Gray. *Bull. Soc. Mycol. Fr.* 91; 313-390. A very important compilation of descriptions of species characteristic of montane and tundra areas, and based on personal collections by the author. Many new taxa are described and montane forms of more familiar lowland species discussed. Unfortunately no line-drawings accompany the data. In French.

Shaffer. F.L. (1975). Some common North American species of *Russula* subsect. *Emeticinae.Nova Nedw., Bein.* 51; 207-237. Seven species two new, in the *Russula emetica* group are described; illustrations of microscopic characters are given.

Grund, D.W. (1979). New and interesting taxa of *Russula* occurring in Washington State. *Mycotaxon* 9; 93-113. Three species, two new, are fully described and supported by line-drawings and b/w photographs.

GASTROID RUSSULALES

Pegler, D.N. and Young, J.W.K. (1972). Gastroid Russulales. *Trans. Brit. Mycol. Soc.* 72; 355-388. Results from an ultra-structure investigation are given. A key to families and genera and descriptions of examples are supplied. B/w micrographs support the account.

STROPHARIACEAE

Genera of this family are very close and are considered together. Quelet placed them all in the genus *Geophila*.

Characters. 1. Gill-attachment. 2. Viscidness of the pileus. 3. Type of veil and if present texture and distribution. 4. Whether basidiomes are aggregated or not.

Hypholoma (= *Naematoloma* fide Singer)

Kuhner, R. (1936). Observations sur le genre *Hypholoma. Bull. Soc. Mycol. Fr.* 52; 9-30. Five species and a possible sixth are keyed-out and discussed. The affinities of the genus are considered. In French.

Smith, A.H. (1951). The North American species of *Naematoloma. Mycologia* 43; 467-521. Twenty-one species are fully described including eight new combinations and one new species. A key is provided and supported with line-drawings and eight b/w photographs. Important contribution to this widely distributed N. Hemisphere group.

Urbonas, V. (1975). On systematics and distribution-area of the fungi of the Strophariacea family in the USSR. 2. *Hypholoma. Lietuvos TSR Mokslu Acad. darbai C.*, 4 (72); 3-17. A key and short descriptions of eighteen species of *Hypholoma* are given. Line-drawings and an English summary are provided. In Russian.

Melanotus

Horak, E. (1977). The genus *Melanotus. Persoonia* 9; 305-327. Illustrated descriptions and a key to twenty-one accepted species are given; five new species are described.

Psilocybe

Singer, R. and Smith, A.H. (1958). Taxonomic monograph of *Psilocybe* sect. *Caerulescentes. Mycologia* 50; 262-303. A standard work; descriptions and key to species are given. Line-drawings are provided. See also *Mycologia* 50; 141-142.

Orton, P.D. (1969). Notes on British Agarics III. *Notes R. Bot. Gdn. Edin.* 29; 75-127. Within this article (pg. 80-82) a key to British *Psilcybe* spp. is given and (pg. 118-122) four species, three new are described. Line-drawings given.

Guzman, G., Varela, L. and Ortiz, J.P. (1977). Les especies no alucinantes del Genero *Psilocybe* concocidas en Mexico. *Bol. Soc. Mex. Mic.* 11; 23-33.

Guzman, G. and Vergeer, P.P. (1978). Index of taxa in the genus *Psilocybe. Mycotaxon* 6; 464-476. (in prep. for a monograph by G. Guzman). Invaluable reference source for original descriptions of species of *Psilocybe.*

Guzman, G. (1978). The species of *Psilocybe* known from Central and S. America. *Mycotaxon* 7; 225-255. Sixty-six species are discussed, fifteen being described as new. No line-drawings offered although tabular material given.

Urbonas, V. (1978). *Lietuvos TSR Mokslu Acad. darbai*; C., 1 (81); 9-18. A key and short descriptions to thirteen species are given. Line-drawings but no English summary. In Russian.

Guzman, G. and Watling, R. (1978). Studies on Australian agarics and boletes I. Species of *Psilocybe. Notes R. Bot. Gdn. Edinb.* 36; 179-210. A full review of all the species recorded in the genus *Psilocybe* for Australia are discussed; three new species are described and illustrated.

Guzman, G. (1979). Observations on the evolution of *Psilocybe* and descriptions of four new hallucinogenic species from Mexican Tropical forests. *Sydowia, Beih.* 8; 168-181. Four new taxa are provided with fully illustrated descriptions.

Guzman, G. and Horak, E. (1979). New species of *Psilocybe* from Papus, New Guinea, New Caledonia and New Zealand. *Sydowia* 31 (1978); 44-54. Six new species of *Psilocybe* are described and illustrated (see shorter article by Guzman with C. Bas on *Psilocybe* (*Persoonia* 9 (1979); 233-238).

Stropharia

Urbanas, V. (1973). On systematics and distribution-area of the fungi of the Strophariaceae family in the USSR. I. *Stropharia-Lietuvos TSR Mokslu Akad. darbai* C, 2 (62); 9-23. A key and short descriptions to fourteen species of *Stropharia* are given. Line-drawings and an English summary are provided. In Russian.

Orten, P.D. (1976). Notes on British Agarics VI. *Notes R. Bot. Gdn. Edin.* 35; 147-153. Key to *Stropharia aeruginosa* grp. offered with full descriptions of two species. No line-drawings given.

Kreisel, H. (1979). Zur taxonomic von *Stropharia aeruginosa* s. lato. *Sydowia, Beih.* 8; 228-232. One new species is described and a key to the *S. aeruginosa* group is provided. No illustrations are offered. In German.

SECOTIOID FUNGI:

Secotiaceous relatives of members of families Agaricaceae to Strophariaceae inclusive.

Singer, R. and Smith, A.H. Studies on Secotiaceous Fungi I-X. (1950-1964).

    I. A monograph of the genus *Thaxterogaster. Brittonia* 10 (1950); 201-216. The genus is discussed and the six species accepted keyed-out. Illustrations are provided.

    II. *Endoptychum depressum. Brittonia* 10 (1950); 216-212. The new species, *E. depressum*, is described and compared with five other accepted species.

III. The genus *Weraroa. Bull. Torrey. Bot. Club* 85 (1950). 324-334. The three species of *Weraroa* then known are described, illustrated and keyed-out.

IV. *Gastroboletus, Truncocolumella* and *Chamonixia. Brittonia* 11 (1959); 205-223. A key is provided to the *Gastroboletus*-series and to species within the three genera accepted; one species is described as new and several new combinations are proposed.

V. *Nivatogastrium* gen. nov. *Brittonia* 11 (1959); 224-228. An illustrated description of the genus and its single species is offered.

VI. *Setchelliogaster. Madrono* 15 (1959); 73-79. Pouzar's genus is expended to include a further species.

VII. *Secotium* and *Neosecotium. Madrono* 5 (1960); 152-158. The genus *Secotium* is restricted to *S. gueinzii*. Descriptions of the species and two taxa in the newly proposed genus *Neosecotium* are given.

VIII. A new genus in the Secotiaceae related to *Gomphidius. Mycologia* 50 (1958); 927-938. A key and illustrated descriptions are given for the three species of the genus. Two are new taxa.

IX. The Astrogastraceous series. *Mem. Torrey Bot. Club.* 21 (1960); 1-112. A key to the genera contained in the series is given and to species within each genus. Seventy-five species in genera are described, many new and several illustrated. A standard work - see also Smith, A.H. (1962). Notes on Astrogastraceous series. *Mycologia* 54; 626-639; Smith, A.H. (1963). New Astrogastracoeus fungi from the Pacific North West. *Mycologia* 55; 421-441.

X. Additional data on *Gastroboletus. Mycologia* 56 (1964); 310-313. As indicated by the title this is a continuation of the 1959 study; see above. Also see: - Smith, A.H. and Reid, D.A. (1962). A new genus of the Sectotiaceae. *Mycologia* 54; 98-104. The genus *Cribbea* is introduced as new, and illustrated descriptions of the three constituent species previously placed in *Secotium* are given.

Singer, R. and Smith, A.H. (1963). A revision of the genus *Thaxterogaster. Madrono* 17; 22-26. A key to the sections of *Thaxterogaster* and to members of those sections (*Thaxterogaster, Microgaster* and *Aporpogaster*) are given. Three species are described as new. Illustrations not supplied. (1963).

Singer, R. (1963). Notes on secotiaceous fungi. *Galeropsis* and *Brauniella. Proc. K. Ned. Akad. Wet. Proc. C,* 66; 106-117. A further expansion of the concept of *Galeropsis* and the consultation of the Galeropsideae. Complements many of the author's earlier papers; supported by line-drawings.

Smith, A.H. (1966). Notes on *Dendrogaster, Gymnoglossum, Protoglossum* and species of *Hymenogaster. Mycologia* 58; 100-124. A key to eleven species of *Hymenogaster* subgenus *Dendrogaster* is given; ten new species are described. A key is also given to twelve N. American species of *Hymenogaster* subgenus *Hymenogaster*; four species are described as new.

Horak, E. and Moser, J. (1966). Fungi Asutroamericani: XII. Studien zur Gattung *Thaxterogaster. Nov. Hedw.* 10; 211-241. Twenty-eight species are described, the majority new; three subgenera are recornised, two new. Line-drawings support the full descriptive data.

Horak, E. (1973). Fungi Agaricini Nova Zelandiae. II. *Thaxterogaster*. in *Nova Hedw. Beih.* 43. Fourteen species are described and illustrated; see pg. 82.

Rhizopogon

Smith, A.H. (1964). *Rhizopogon*; a curious genus of false truffles. *Mich. Bot.* 3; 13-19. Introductory material for forthcoming monograph. Key to the six accepted sections of the genus, five new are given; two new species are also described as new.

Smith, A.H. and Zeller, S.M. (1966). A Preliminary account of the North American species of *Rhizopogon*. *Mem. N. Y. Bot. Gdn.* 14 (2); 1-178. One hundred and thirty-seven species are recognized, numerous being new species. Keys and descriptions are given to the species and sections; sporograms are given most taxa and some photographic and coloured illustrations of basidiomes supplied. See also Further Studies on *Rhizopogon* I. *J. Elisha Mitchell Scienc. Soc.* 84 (1968); 274-280. Nine species are considered, five of which are new; the latter are fully described.

Harrison, K.A. and Smith, A.H. (1968). Some new species and distribution records of *Rhizopogon* in North America. *Can. J. Bot.* 46; 881-899. Twelve new species are described from British Columbia and New Mexico. Additional information is presented on four recently described species and ranges of twenty-nine other species extended.

TRICHOLOMATACEAE

The genera within this family are often very close, indeed many authorities do not accept the fragmentation of the traditional and familiar genera. See Bon, M. (1972). Etudes microscopiques. *Docums. Mycol.* 2 (5); 29-32.

Characters.   1. Presence of absence of ring. 2. Outer layer of stipe-cartilagineous or fibrous, wiry or not. 3. Extent of pileus striation. 4. Attachment of gills. 5. Viscidness or scaliness or pileus. 6. Whether basidiomes are solitary, or in troops, or caespitose. 7. Habitat preference; in case of small species record the actual debris on which they are found (or failing this preserve with the specimen a piece for later identification). (includes *Rhodocybe, Lepista* and *clitopilus* etc.)

GENERAL

Singer, R. (1964). *Oudemansiellineae, Macrocystidineae,* and *Pseudohiatulineae* in S. America. *Darwiniana* 13; 145-190.

Lange, M. and Silvertsen, S. (1966). Some species of *Lyophyllum, Rhodocybe* and *Fayodia* with rough spores: Nomenclature and taxonomic position. *Bot. Tidssk.* 62; 197-211. Descriptions of a number of rough-spored European omphaloid and collybioid species are given. No illustrations are provided.

Gulden, G. (1966). Cone-inhabiting agarics with special reference to Norwegian material. *Nytt Mag. Bot.* 13; 39-55. Five species are described and illustrated, three in *Strobilurus* and one each in *Mycena* and *Baeospora*.

Singer, R. (1970). *Flora Neotropica* 3; 1-84. Covers four taxa or *Omphalina*,

52

three of *Callistosporium*, one of *Arthrosporella* with its conidial state, five
of *Armillariella*, six of *Lactocollybia*, two of *Macrocystidia*, twenty of
Gerronema, three of *Pleurocollybia*, and one of *Lulesia*.

Singer, R. and Clemenson, H. (1972). Notes on some leucosporous and
rhodosporus European agarics. *Nova Hedw.* 23; 305-344. One species each
in *Geopetalum*, *Hygrocybe*, *Armillaria*, *Gerronema*, *Crinipellis*, *Hydropus* and *Squamanita* are described as are two each in *Lepist*, *Omphalina*,
Pluteus and *Cystolepiota* and three in *Melanoleuca*. Twelve figures of
line-drawings and seven plates of b/w photographs of macroscopic structures are supplied.

Binyamini, N. (1973). White-spored agarics new to Israel. *Israel J. Bot.* 22;
33-46. Eighteen taxa in *Cellybia*, *Mycena*, *Clitocybe*, *Melanoleuca* and
Pleurotus are discussed, all but one being new records. Seven species are
illustrated in b/w photographs; one plate of line-drawings is also
provided.

Bon, M. (1978). Tricholomataceae de France et d' Europe occidentale.
Docums. Mycol. 33; 1-79. Well documented contribution to sub-family
Leucopaxilloideae with descriptive keys, line-drawings to *Leucopaxillus*,
Porpoloma and *Melanoleuca*, species of *Cantharellula*, *Pseudoclitocybe*,
Pseudomphalina and *Clitocybula* are also keyed out. Tribe Biannulariae,
consisting of *Catatheasma* and *Floccularia* (= *Armillaria*) are also
described.

Bigelow, H. (1975). Studies in the Tricholomataceae. *Hygrophoropsis*, *Cantharellula*, *Myxomphalina*, *Omphaliaster*. *Nova Hedw.*, *Beih.* 51; 61-77.
Seven species are provided with descriptions and are keyed-out. Three
plates of b/w photographs are given.

Anthracophyllum

Singer, R. (1977). Keys to species. *Sydowia* 30; 211. A key to the seven species
known is given.

Armillaria S. lato. Includes various elements now placed in *Tricholoma*,
Pleurotus etc.

Kauffman, C.H. (1923). The genus *Armillaria* in U.S. and its relationships.
Paper. Mich. Acad. Sci. 2; 53-67. Keys and observations are provided;
supported by five b/w photographs. See Hobson, H.H. (1940). The genus
Armillaria in Western Washington *Mycologia* 32; 776-790. Includes key
to the genus and three b/w photographs.

Thiers, H.D. and Sundberg, W.J. (1976). *Armillaria* in the Western United
states including a new species from California. *Madrono* 23; 448-453. A
key is provided to ten species with full descriptions of a single new species.
No illustrations provided; includes members of *Tricholoma* s. st. and
Floccularia.

Armillaria - a confused concept in many accounts; *Armillaria* as here
adopted covers *Armillariella*.

Singer, R. (1956). The *Armillaria mellea* group. *Lloydia* 19; 176-187. A good
introduction with descriptions of five species, three of which are new; a
new variety is also published and a key provided. B/w photographs of two
species.

Singer, R. (1970). *Armillaria mellea. Schweiz Z. Pilzk.* 48; 25-29. Examination
of the problems within the group. In German.

Singer, R. (1970). A monograph of the subtribe Omphalinae. *Flora
Neotropica* 3; 1-84. *Armillariella* is covered in first 16 pages with key and
descriptions of five species, one new; one subspecies is also described as
new and one new combination made. The closely related *Lulesia* and
Arthrosporella formerly placed in *Armillaria* are also described.

Singer R. (1956). The *Armillaria mellea* group. *Lloydia* 19; 176-187. A good
introduction with descriptions of five species, three of which are new; and
a new variety is also published and a key provided. B/w photographs of
two species.

Singer, R. (1970). *Armillaria mellea. Schweiz Z. Pilzk.* 48; 25-29. Examination
of the problems within the group. In German.

Singer, R. (1970). A monograph of the subtribe Omphalinae. *Flora Neotropica*
3; 1-84. *Armillariella* is covered in first 16 pages with key and descriptions of five species, one new, one subspecies also described as new and one
new combination made. The closely related *Lulesia* and *Arthrosporella*
formerly placed in *Armillaria* are also described.

Romagnesi, H. (1970 and 1973). Observations sur les *Armillariella*. I. *Bull.
Soc. Mycol. Fr.* 86; 257-265. II. Ibid 89; 195-206. *A. ostoyae* and *A. obscura*
are described as new species in I. Detailed descriptions of a typical 'form'
of *A.mellea* and of *A.bulbosa* are given in II.

Singer, R. (1977). Key to species of *Armillariella. Sydowia* 30; 211-216. A key
is provided to all the species accepted by the author.

Callistosporium

Singer, R. (1977). Key to species. *Sydowia* 30; 261-262. A key to the eleven
species known is given, two are described as new.

Calocybe

Singer, R. (1977). *Sydowia* 30. Key to the twenty-four species known is given;
five new species are described.

*Cantharellula

Singer, R. (1977). Key to species of *Cantharellula. Sydowia* 30; 279. A key is
provided to all known species.

Cantharocybe

Bigelow, H. and Smith, A.H. (1973). *Cantharocybe*, a new genus of Agaricales.
Mycologia 65; 485-486. A full description with one figure of line-drawings
is given.

Catathelasma

Singer, R. (1979). Key to species of *Catathelasma. Sydowia* 31; (1978); 193-194.
The five known species are keyed-out.

Chaetocalathus See Singer under *Crinipellis* below.

Singer, R. (1979). Key to species of *Chaetocalathus, Sydowia* 31 (1978); 194-
196. Key to all known species; one new species is described. Supersedes
earlier account by Singer.

54

Clitocybe

Raithelhuber, J. (1970). Various articles in *Matrodiana* which appears to be a
journal restricted to studies of Tricholomataceae, especially *Clitocybe* s.
lato.

Murrill, W.A. (1915). The genus *Clitocybe* in N. America. *Mycologia* 7; 256-
283. See Kauffman below, although no reference to Murrill's article is
made therein. Several new combinations and new names are made and
three new species from Eastern N. America proposed. B/w photographs
support the text.

Kauffman, C.H. (1927). The genus *Clitocybe* in the United States with a
critical study of all north temperate species. *Pap. Mich. Acad. Sci.* 8;
153-204. The article is divided into an introductory account of the genus, a
key to species of the North Temperate zone divided into five groups, and
descriptions of twelve new taxa. The final section comments on some
further species with one new name and one new combination and an index
of synonyms and excluded or doubtful species.

Bigelow, H.E. (1958). New species and varieties of *Clitocybe* from Michigan.
Mycologia 50; 37-51. Eleven new species are described; two b/w photographs are supplied.

Bigelow, H.E. and Hesler, L.R. (1960). *Clitocybe* in Tennessee and N. Carolina
J. Elisha Mitchell Scient. Soc. 76; 155-167. Twelve species are illustrated
by b/w photographs. Unfortunately no full descriptions are given.

Bigelow, H.E. and Smith, A.H. (1962). *Clitocybe* species from the Western U.
States. *Mycologia* 54; 498-515. Ten species, three are described; four b/w
photographs are supplied.

Bigelow, H.E. (1965). The genus *Clitocybe* in N. America: Section *Clitocybe*.
Lloydia 28; 139-180. A key is supplied to thirty species, three of which are
new; one taxon is supplied with a new name. Twenty-six b/w photographs
and a list of excluded species are supplied. No line-drawings of micro-
characters are given.

Harmaja, H. (1969). The genus *Clitocybe* (Agaricales) in Fennoscandia.
Karstenia 10; 1-120. Keys and full descriptions are given to all species
recorded for Fennoscandia; discussion on many is offered and supported
by ecological data. Line-drawings of habit and basidiospore outlines (on
which emphasis is placed, 3 major plates) and maps of distribution support the data. The sixteen b/w photographs of dried basidiodomes
included are of limited use. Fourteen new species and several new sections and other taxa are proposed but see Harmaja's concept of *Clitocybe*
in *Karstenia* 14 (1974); 82-92.

Harmaja, H. (1970-1979). Type studies in *Clitocybe* I. *Karstenia* 11; 35-40.
Seventeen species and one form are scrutinized in *Clitocybe*, and one in
Omphalina. II. Ibid 15; 16-18. Holotypes of seven taxa are examined. III.
Ibid 19; 22-24. Six holotypes are discussed and connections between
Lyophlleae and *Rhodocybe* noted. IV. Ibid 19; 50-51. Holotypes of four N.
American taxa are considered and *C. schumannii* reconsidered. No illus-
trations are provided.

Kuhner, R. and Lamoure, D. (1972). Agaricales de la zone Alpine, *Clitocybe*.
Trav. Sc. Parc. Nat. Vanoise 2; 107-152. Detailed study of twenty-three
species of *Clitocybe* including six new species from alpine zones is given

and supported by sporograms. In French.

Bigelow, H.E. (1976). Studies on some lignicolous *Clitocybe. Mem. N.Y. Bot. Gdn.* 28; 9-15. Eight species, two new, are considered and described. Two species are transferred to Hypsizygus. No line-drawings are given.

Singer, R. (1979). Key to species of *Clitocybe. Sydowia* 31 (1978); 199-233. Key to all species accepted by author; includes descriptions of four new species and one species ad. int. and one new combination.

Clitocybula

Romagnesi, H. (1968). Sur un *Collybia* a spores amyloides. *Collnea Bot. (Barsinone)* 7; 1083-90. One new variety is described and illustrated.

Bigelow, H.E. (1973). The genus *Clitocybula. Mycologia* 65; 1101-1116. A monographic treatment of the N. American species. Six species are described, illustrated and keyed-out.

Singer, R. (1979). Key to species of *Clitocybula. Sydowia* 31; 233-234. Keys to all species known, with description of a single new species.

Clitopilus

Singer, R. (1946). The Boletineae of Florida IV. Lamellate families. *Farlowia* 2 (4); 531-566. Members of the genus *Clitopilus*, placed in the Jugasporaceae, are keyed-out with sections of the genus; notes are included on extra-limital species. Part of a wide-ranging review of importance outside area defined. Two new sections and one new species are described in addition to four additional species.

Singer, R. (1979). Key to species of *Clitopilus. Sydowia* 31 (1978); 235-237. A key to the species accepted by the author; one new combination is made.

Collybia s. lato.

Lennox, J.W. (1979). Collybioid genera in the Pacific North West. *Mycotaxon* 9; 117-231. A key to the eleven genera considered is given and each genus is discussed in turn. *Microcollybia* and *Caulorhiza* are described as new. Thirty-four species and subspecies are included, four of which are new, and eighteen proposed as new combinations.

Crinipellis

Singer, R. (1942). A monographic treatment of the genera *Crinipellis* and *Chaetocalathus. Lilloa* 8; 441-534. The characters defining the species of *Crinipellis* are given with key to the thirty-three accepted species. A second key based on geography and substrate is given. Full descriptions are offered with two plates of line-drawings which although small are useful. The eleven species of *Chaetocalathus* are treated similarly. An index is given. Several new taxa and combinations are proposed.

Singer, R. (1976). *Flora Neotropica* 17; 9-58. Part of a much bigger work. Keys and descriptions to forty-one species of *Crinipellis* are given including the cause of "Witch's broom of Cocoa' see Pegler D.N. (1978). *Kew Bull.* 32; 731-736, for further information on the latter including descriptions of further infraspecific taxa.

56

Cystoderma

Smith, A.H. and Singer, R. (1945). A monograph on the genus *Cystoderma*.
Pap. Mich. Acad. Sci. 30; 71-124. The first synthesis of the generic limita-
tion and taxa involved. Full descriptions of fourteen species given and
variation within the taxa discussed. B/w photographs support the text.

Locquin, M. (1951). Le genre *Cystoderma* Fayod. Etude de especes francaise.
Bull. Soc. Mycol. Fr. 67; 65-80. A key to sections and all species recorded
for France is given and supported by descriptions of six species; accompanied by line-drawings. In French.

Thoen, D. (1967). Les Cystoderma (Tricholomataceae). *Naturalistes belg.* 48;
285-297. Keys are given to Belgian and European species with short
descriptions and distribution-maps for four species; *Phaeolepiota aurea* is
included. In French.

Kuhner, R. (1969). *Cystoderma amianthinum* var. *longisporum* et var.
sublongisporum. Bull. Soc. Linn. Lyon 38; 178-188. Full discussion on this
polymorphic species. In French.

Thoen, D. (1969). Le genre *Cystoderma* (Tricholomataceae) en Afrique
centrale. *Bull. Jard. Bot. Nat. Belg.* 39; 183-190. The tropical species of the
genus are reviewed and two species are described as new; one new combination is made. Line-drawings are provided. In French.

Thoen, D. and Heinemann, P. (1973). *Flore Illust. Champignons Afrique Cent.*
Fasc. 2. see pg. 84.

Heinemann, P. and Thoen, D. (1973). Observations sur le genre *Cystoderma*.
Bull. Soc. Mycol. Fr. 89; 5-34. A review is given of the characters used in
the identification of species. Two new species are described and keys to all
recognized species given. A useful tabulation of specific names with notes
is given. Line-drawings, one b/w photograph and one coloured plate
supports the text. In French.

Harmaja, H. (1979). Studies in the genus *Cystoderma. Karstenia* 19; 25-29.
Four species, one new are considered; one b/w photograph is provided.

Fayodia

Bigelow, H. (1979). Notes on *Fayodia* s. lato. *Mycotaxon* 9; 38-47. One new
species and the genus *Stachyomphalina* are proposed, the latter for
Clitocybe striatula. Full descriptions and micrographs supplied.

Gerronema

Singer, R. (1964). Die Gattung *Gerronema. Nova Hedw.* 7; 53-92. Keys to and
descriptions of thirty species are given, five other taxa are mentioned. No
line-drawings are given. In German.

Haasiella

Kotlaba, F. and Pouzar, Z. (1966). *Haasiella* a new agaric genus. *Ceska Mykol.*
20; 135-146. Descriptions of the genus and type species are supported by
line-drawings of basidia and spores, b/w photographs and one coloured
plate. In English; Czech summary.

Hydropus

Moser, M. (1968). Uber eine neue art aus der Gattung *Hydropus* (Kuhn.)
Singer. *Z. Pilzk.* 34; 145-151. One new species is described and a key to

European species provided. Line-drawings are given. In German.

Laccaria

Singer, R. and Moser, M. (1965). Forest Mycology and Forest communities in S. America I. *Mycopath. Mycol. Appl.* 26; 191-192. A very basic review, with collaborators but includes key to species of *Laccaria* found in S. America.

Heinemann, P. (1966). *Flore Icon. Champignons Congo.* Fasc. 15. see pg. 83.

Singer, R. (1967). Notes sur le genre *Laccaria. Bull. Soc. Mycol. Fr.* 83; 104-123. Keys are provided for *L. laccata* and *L. tetraspora* and the varieties known to the author. Tabular data for *L. tetraspora* and *L. ohiensis* and notes on four additional species are given.

McNabb, R. (1972). Tricholomataceae of New Zealand I, *Laccaria. N. Z. J. Bot.* 10; 461-484. Twelve species, two new, and sub-specific taxa of *Laccaria* occuring in N. Zealand are described and illustrated. Two new varieties are proposed and all taxa are illustrated by line-drawings of microscopic characters. A key to the species is given and doubtful records discussed.

Aguirre-Acosta, E. and Perez-Silva, E. (1978). Description de algunas especies del genero *Laccaria.* (Agaricales) 1. *Laccaria. Bol. Soc. Mex. Mic.* 21; 33-58.

*Lepista

Bigelow, H. and Smith, A.H. (1-69). The status of *Lepista* - a new section of *Clitocybe. Brittonia* 21; 144-177. The genus *Lepista* is discussed and accommodated within *Clitocybe.* Two new species, two new varieties and a new section are proposed; several new combinations are made. Eighteen species and three varieties are treated; a key to taxa and complete descriptions of such species are given. Sixteen b/w photographs are given but no line-drawings offered.

Kuhner, R. (1976). Agaricales de la zone Alpine, Lepistees. *Bull. Soc. Mycol. Fr.* 92; 5-32. Three species of *Lepista* and one of *Ripartites* are described following a nomenclatorial and taxonomic study of the group. In French.

*Leucopaxillus

Singer, R. and Smith, A.H. (1943). A monograph of the genus *Leucopaxillus. Pap. Mich. Acad. Sci.* 28; 85-132. Keys are given to twelve species and to varieties within these species. Eight b/w photographs are provided along with one line-drawing of basidiospores and pileipellial hyphae. Full descriptions of all taxa are given. (Additional notes on the genus *Leucopaxillus. Mycologia* 39; 725-736. Additional notes to the 1943 publication including a new species and a new variety).

Marasmiellus

Singer, R. (1973). A monograph of Neotropical species of *Marasmiellus. Nova Hedw., Beih.* 44; 1-340. One hundred and thirty-four species are accepted and twelve extra-limital species discussed and thirty-four taxa excluded. Twelve pages of Latin diagnoses are given and one hundred and twelve line-drawings offered although some are very sketchy. A list of hosts and epithets considered is provided. A tremendous work and a noteworthy contribution with useful key to species and other taxa.

Singer, R. (1975). The neotropical species of *Campanella* and *Aphyllotus* with notes on some species of *Marasmiellus*. *Nova. Hedw.* 26; 847-893. A monographic treatment of the fifteen species, many new, known to occur in the neotropics is given with descriptions and keys, and with the redescription of *Aphyllotus* and its single species. In addition eight new species and one combination are proposed in *Marasmiellus* as a sequence to the 1973 article by the same author. The line-drawings supplied are rather sketchy; one coloured plate is provided.

Marasmius

Kuhner, R. (1933). Etudes sur le genre *Marasmius*. *Botaniste* 25; 57-114. Includes studies on the development of *M. rotula*, cytology of the hymenium of *M. androsaceus* and *M. rotula*, microchemistry of cell-walls of *Marasmius* spp. and classification of members of the genus. A key is offered and full descriptions of several taxa given.

Singer, R. (1936). Studien zur Systematik die Basidiomyceten I. Abgrenzung zwisehen *Collybia* and *Marasmius*. *Beih. Bot. Zbl.* Abt. B 56; 157-163. Keys to sections of *Marasmius* and its relationships to other marasmoid agarics offered.

Petch, T. (1948). A review of Ceylon *Marasmius*. *Trans. Br. Mycol. Soc.* 31; 19-44. Fifty-four species, two new, in the genus *Marasmius* s. lato are described from Sri Lanka. The descriptive data is accompanied by 3 beautiful coloured plates depicting forty-seven illustrations. No line-drawings of spores etc. given. A few species dubiae and excludendae are given.

Singer, R. (1958). Studies towards a monograph of S. American species of *Marasmius*. *Sydowia* 12; 54-158. A survey of the sections is given with descriptive key to almost ninety taxa. Ten new species, several new names and new varieties are proposed, as well as new combinations; some line-drawings are offered. Smaller keys to varieties and closely related taxa in critical groups support the descriptive data.

Singer, R. (1964). The genus *Marasmius* in S. America. *Sydowia* 18; 106-578. Part of the monographic studies on S. American Basidiomycetes especially those from the east slopes of the Andes and Brazil. One hundred and forty-three species, many new, are accepted and supported by descriptive data and twenty-two figures of line-drawings given. Thirteen pages of Latin diagnosis are given, as well as a supplement to six further species and index of taxa.

Singer, R. (1965). *Flore Icon. Champignons Congo.* Fasc. 14. see pg. 83.

Gilliam, M.S. (1975). New North American species of *Marasmius*. *Mycologia* 67; 817-844. Nine new species are fully described from the North Eastern United States, three species in section *Chordales*, four in section *Sicci*, and one each in sections *Epiphylli* and *Androsacei*. Taxa are fully documented in nine plates of line-drawings of microscopic features.

Gilliam, M.S. (1975). *Marasmius* section *Chordales* in the N.E. United States and adjacent Canada. *Contr. Univ. Mich. Herb.* 11; 25-40. A key is provided to six species fully described and documented with line-drawings and b/w photographs.

Singer, R. (1976). Marasmieae *Flora Neotropica* 17; 1-347. This contribution includes *Gloiocephala*, (*Marasmius* pg. 61-284), *Physalacria*, *Epienaphus*, *Manuripia*, *Humenogloea*, and *Rimachia* pg. 284-315 and *(Crinipellis*, *Chaetocalathus*) *Lachnella*, *Flagelloscypha* and *Amyflagellula* pg. 9-61. Keys, descriptions and discussion on both agaricoid and reduced members. Line-drawings are supplied and many new taxa proposed: see *Crinipellis* and *Chaetocalathus* above. Two hundred and thirty-three species are considered along with a list of incompletely known taxa given.

Smith, A.H. (1979). The stirps *Cohaerens* of *Marasmius*. *Mycotaxon* 9; 341-347. A key and one new species are given; b/w photographs are supplied.

Melanoleuca

Singer, R. (1935). Etudes systematique sur les *Melanoleuca* d'Europe et cle des especes observees en Catalogne. *Cavanillesia* 7; 122-132. A major contribution to an understanding of the genus and placing the definition of taxa within the genus on a modern footing.

Metrod, G. (1942). Sur le genre *Melanoleuca*. *Rev. Mycol.* 7; 89-96. Four species are considered in this short contribution; line-drawings of basidiomes and micro-characters are given. In French.

Metrod, G. (1948). Essai sur le genre *Melanoleuca* Pat. *Bull. Mycol. Fr.* 65; 141-165. A continuation of the 1942 study with full introductory discussion and notes on rejected taxa; one place is provided. A key and full descriptions of twenty-one species are given; supported by line-drawings. In French.

Gilliam, L.S. and Miller, O.K. (1977). A study of the boreal alpine and arctic species of *Melanoleuca*. *Mycologia* 69; 927-951. Ten species, five new are described and a key presented. Line-drawings and b/w photographs support the text which includes ecological data.

Kuhner, R. (1978). Agaricales de la zone Alpine; *Melanoleuca*. *Bull. Soc. Linn. Lyon.* 47; 12-52. Eight species are described with several varieties; one new species and two new varieties are proposed and variation in *M. cognata* discussed to include new combinations. Line-drawings offered. In French.

Mycena see Charbonnel, J. (1977). Etudes microscopique...*Docums. Mycol.* 7 (26) 1-70.

Oort, A.J.P. (1928). De Nederlandsche Mycenas. *Meded. Ned. Mycol. Vereen* 16-17; 163-183. A very useful contribution.

Kuhner, R. (1938). *Le Genre Mycena. Encycl. Mycol.* 10; Paris 710 pp. A standard work covering taxa other than in East Continental Europe. Species are described in full with numerous supporting line-drawings. A general comprehensive introduction to the genus is offered. A standard work which has never been superseded. In French.

Smith, A.H. (1947). *North American Species of Mycena*. Ann Arbor, pp. 521. Complementary to Kuhner's work but more restricted to N. America; never-the-less important outside the limits of that continent. Many new taxa described. Full descriptions supported by numerous line-drawings of basidia, spores and cystidia, and b/w photographs. A reprint of this standard work is now available as *Bibl. Mycol.* 31 (1971). Includes 'Studies in *Mycena*'. *Mycologia* 27 (1936)-31 (1939).

Pearson, A.A. (1955). *Mycena. Naturalist* (1955); 41-63. Manuscript completed by R.W.G. Dennis in the familiar tabular form. Keys and descriptions offered to a-1 species then recorded for British Isles. A useful check-list of names is also given. Re-issued with commentary and introduction by R. Watling in 1978.

Metrod, G. (1959). Les Mycenes de Madagascar. *Prod. Fl. Mycol. Madagascar* No. 3. One of a series of extensive and important studies from this part of the world (see pg. 119). Eighty-seven species, most new are accepted in three genera; two genera are new. Ten pages of Latin validations are given and several line-drawings supplied. In French.

Kuhner, R. and Valla, G. (1972). Contributions a la connaissance des especes blanches a spores non amyloids du genre *Mycena. Trav. lab. Jaysinia* 4; 25-71. Twenty-four species or varieties are fully described, about half being described for the first time. Ten plates of line-drawings support the descriptive data. In French.

Horak, E. (1978). *Mycena rorida* and related species from the southern Hemisphere. *Ber. Schweiz. Bot. Ges.* 88; 20-29. A key is offered to six species. Three new species are described and illustrated with line-drawings. Their luminescence and taxonomic relationships with *M. rorida* are discussed. In English.

Omphalina

Cejp. K. (1935). *Omphalina; Atlas des Champignons de l'Europe* 4. Prague, 152 pp. One hundred and eight European taxa are covered; *Omphalina* is taken in the wide-sense. Vol. 4B covers ten species of *Delicatula*, some formerly placed in *Omphalina*. A full checklist of names is given along with line-drawings and b/w photographs and key.

Bigelow, H.E. (1970). *Omphalina* in North America. *Mycologia* 62; 1-32. The genus is emended for species with yellow to orange pigments. Fourteen taxa are recognized for N. America and are described and keyed-out. Ten b/w photographs support the data and a check list of '*Omphalina*'species given. No line-drawings offered.

Lamoure, D. (1974-75). Agaricales de la zone Alpine, *Omphalina. Trav. Sc. Parc Nat. Vanoise* 5; 149-164 and 6; 153-166. Two parts covering twelve species, five new, six each in brown and dark grey-brown groups; supported by line-drawings and sporograms. Full descriptions given in French

Omphalotus

Bigelow, H.E. and Miller, O.K. and Thiers, H.D. (1976). A new species of *Omphalotus. Mycotaxon* 3; 363-372. Line-drawings and b/w photographs support the recognition of a new species within the genus. Discussion is offered.

Panellus

Miller, O.K. (1970). The genus *Panellus* in N. America. *Mich. Bot.* 9; 17-30. (see later addition of *Dictyopanus* by Miller, O.K. and Burdsall, H. see pg. 65.)

Pseudobaeospora

Singer, R. (1963). The delimitation of the genus *Pseudobaeospora. Mycologia*

60; 13-17. Two new species are described, discussed and illustrated in line-drawings.

*Rhodocybe

Kuhner, R. (1971). Agaricales de la zone Alpine, *Rhodocybe. Bull. Soc. Mycol. Fr.* 87; 15-23. Three species are considered. No line-drawings provided. In French.

Horak, E. (1979). Notes on *Rhodocybe* Maire. *Sydowia* 31; (1978). 38-80. Descriptive data supported by line-drawings covering twenty-four taxa and one excluded species are given. Species from Australia, Papua, Sabah Java, Singapore, Sri Lanka, Switzerland, England, Norway and Greenland are considered.

Horak, E. (1979). Fungi Agaricini Nova Zealandiae VII. *Rhodocybe* Maire. *N. Z. J. Bot.* 17; 275-285. Nine species, four new, are provided with illustrated descriptions.

*Ripartites

Hujsman, H.S.C. (1960). Observations sur le genre *Ripartites. Persoonia* 1; 335-339. Two species, one new are described and figured. In French.

Singerella

Harmaja, H. (1974). *Singerella*, a separate genus for *Clitocybe hydrogramma. Karstenia* 14; 113-116. Full descriptions of species and genus are given.

Squamanita

Bas, C. (1965). The genus *Squamanita. Persoonia* 3; 331-359. An emended description of the genus is given and a key to seven taxa, one new and two provisional species are offered. The relationships of the genus are discussed. Line-drawings support the full descriptions of the taxa.

Strobilurus

Well, V.C. and Kempton, P.E. (1971). Studies on the fleshy fungi of Alaska. *Strobilurus* with notes on extra-limital species. *Mycologia* 63; 370-379. Three new species are described and three new combinations made. Full descriptions are given to the six species keyed-out. Line-drawings offered.

Redhead, S.A. (1980). The genus *Strobilurus* in Canada with notes on extra-limital species. *Can. J. Bot.* 58; 68-83. Three species are described and others discussed. A key to all known species is given.

Termitomyces

Heim, R. (1943). Nouvelles etudes descriptives sur les Agarics termitophiles d'Afrique tropicale. *Arch. Mus. Natn. Hist. Nat.* 18; 107-166. Superseded by Heim (1977). *Termites et Champignons*, Paris, 180 pp.

Heim, R. (1958). see *Flora Icon Champignons du Congo.* Fasc. 7 see pg. 83.

Natrajan, K. (1975). South Indian Agarics I. *Termitomyces. Kavaka* 3; 63-66. Four species, one new, are described; line-drawings are supplied (also see *Curr. Sci.* 46; 679-680 for additional species.

62

Tricholoma s. stricto

Horak, E. (1964). Fungi Austroamericani. *Sydowia* 18; 153-163. Twelve spe-
    cies, seven new, are considered and twenty-five plates of line-drawings
    supplied. In German.
Bon, M. (1967-70). Revision des Tricholomes I. *Bull. Soc. Mycol. Fr.* 83;
    324-335. sect. *Albobrunnea*: visquex, II. Ibid 85; 475-492. sect. *Albobrun-
    nea*: secs, and sect. *Inamoena*. III. Ibid. 86; 755-763. sect. *Sejuncta* with
    one coloured plate. Descriptions and keys are provided. In French.
Huijsman, H.S.C. (1968). Observations sur les Tricholomataceae (I). Le
    groupe *Tricholoma terreum* s. stricto. *Schweiz Z. Pilzk.* 9; 143-152. Com-
    parative data and descriptions with one line-drawing are offered for
    members of this group. In French, summary in German.
Bon, M. (1970). Revision des Tricholomes *Bull. Soc. Mycol. Fr.* 39; 755-763. The
    T. sejuncta group of eight species is discussed. One coloured plate is given,
    see also *Docums. Mycol.* 83; 324-335. In French.
Singer, R. (1971). Neue Arten von Agaricales. *Schweiz Z. Pilzk.* 49; 118-128.
    Nine new species are described and figured. One each in *Aeruginospora,
    Dermoloma, Oudemansiella, Pholiota* and *Galerina* and two in both
    Gerronema and *Fayodia*.
Gulden, G. (1972). *Musseronflora slekten Tricholoma sensu lato.* 96 pp., Uni-
    versiteforlagit, Oslo, Bergen and Troms. A well illustrated account of
    those species of *Tricholoma* s. lato recorded for Norway.
Bon, M. (1974-78). Tricholomes de France et d'Europe occidentale I. *Docums.
    Mycol.* 3 (12); 1-53. General account with seven plates of line-drawings. II.
    Ibid 4 (14); 55-110. *Tricholoma* sects. *Saponaceum, Inamoena, Padinicu-
    tus, Atrosquamosa* p.p.; supported by eight plates of line-drawings and
    one coloured plate. III. Ibid 5 (18); 111-164 sect. *Atrosquamosa (terreum*
    grp.), *Equestria* (subsect. *Albata*) with fifteen plates of line-drawings. IV.
    Ibid 6; (22-23) 165-304 sects. *Sejuncta, Imbricata* and *Albobrunnea* with
    thirty-four line-drawings and two coloured plates. In French.
Bigelow, H.E. (1979). A contribution to *Tricholoma. Sydowia, Beih.* 8; 54-62.
    Four taxa and five additional type collections are provided with descrip-
    tive material; one b/w photograph is supplied.
Smith, A.H. (1979). The stirps *Caligata* of *Armillaria* in N. America. *Sydo-
    wia, Beih.* 8; 368-377. Four species, one new, are described. A key is
    provided to the group and two new varieties of *A. caligata* offered. One
    b/w photograph is supplied.

Tricholomopsis

Smith, A.H. (1960). *Tricholomopsis* in the W. Hemisphere. *Brittonia* 12;
    41-76; A key to nineteen species is given along with descriptions of each
    taxon. Notes are given on excluded species. Seven line-drawings of
    microscopic characters are given and several b/w photographs provided.
    A standard publication.

Xeromphalina

Singer, R. (1965). Monograph of South American Basidiomycetes, *Xerom-
    phalina. Bot. Soc. Arg. Bot.* 10; 302-310. A useful account of this genus in
    S. America.

Miller, O.K. (1968). Revision of the genus *Xeromphalina*. *Mycologia* 60; 156-188. Three sections (one new), four subsections (one new) are erected to accommodate the twelve species and varieties from Europe and N. America. One new species and one new variety are described and a key offered for the identification of all taxa. Full descriptions are given and supported by ecological data and illustrations of important structures. B/w photographs of three species are offered.

CUPULOID ELEMENTS

Calyptella
Singer, R. (1977). Key to species. *Sydowia* 30 (1978); 270-271. Keys to the known species in the genus are given and distributed within two sections; one new species is described.

Cellypha
Singer, R. (1979). Key to species. *Sydowia* 31 (1978); 194. Two species are recognized and separated out in a key.

Chromocyphela
Singer, R. (1979). Key to species. *Sydowia* 31 (1978); Two species are recognized and separated out.

Favolaschia
Singer, R. (1974). A monograph of *Favolaschia Nova Hedw., Beih.* 50; 1-108. The genus is placed in the Aphyllophorales (= Polyporales) in its own family close to *Aleurodiscus* and *Gloeosoma*. Fifty one species and several subspecies are recognized and described. Twenty-eight species are not known sufficiently but may well belong to the genus and are keyed-out in a separate artificial key. Line-drawings, several uniformative, and keys support the text.

Flagelloscypha
Aeger, R. (1975). Studien en cyphelloiden Basidiomyceten. *Sydowia* 27; 131-265. A masterly prepared monographic treatment of this small although ever enlarging genus of cupuloid basidiomycetes. Good introduction and discussion of characters, many analyzed statistically, followed by descriptions of all accepted species and discussions on those related or excluded ad int. Many new taxa are proposed. Beautiful line-drawings support the text. In German with key and summary in English.

Gloiocephala
Bas, C. (1961). The genus *Gloiocephala* Massee in Europe. *Persoonia* 2; 77-89. Full descriptions of three European species one un-named are given; supported by line-drawings of both basidiomes and microscopic characters. The merits of accepting the genus are discussed. See also Singer, R. (1952). Le genre *Gloiocephala*. *Rev. Mycol.* 17; 161-164. *G. epiphylla* is described and discussed. In French.

POROID ELEMENTS

Singer, R. (1945). The *Laschia*--complex. *Lloydia* 8; 170-230. An important
paper in the unravelling of the poroid agarics and superficially similar
fleshy fungi. Supported by full descriptions and b/w photographs but
illustrations are of little real use. Good bibliography and index to all
names which may be construed to have been used in *Laschia* or one of its
superficially similar genera. Twenty-five new combinations are made
and two new species proposed.

Heim, R. (1945). Les Agarics tropicaux a hymenium tubule. *Rev. Mycol.* 10;
3-61. Examples from all the following genera are described and illustrated both by line-drawings and b/w photographs: *Dictyopanus*, *Filo-
boletus*, *Laschia*, *Mycena*, *Mycenoporella*, *Mycomedusa*, *Phaeomycena*,
Phlebomycena and *Poromycena*. Areas of the world include Madagascar,
Ivory Coast, New Zealand and tropical French colonies. In French.

Singer, R. (1974). A monograph of *Favolaschia. Nova Hedw., Beih.* 50; 1-108.
see above, pg. 63.

PLEUROTOID TRICHOLOMATACEAE AND PLEUROTACEAE

Characters. 1. Presence or absence of stipe and if present its position. 2. Presence of velar remnants. 3. Spore-print colour. (4. Presence or absence of thick-walled hyphae s.m.).

Pilat, A. (1928). *Pleurotus: Atlas des Champignons de l'Europe.* II. Prague, 193
pp. The genus is taken in the widest sense and includes a whole range of
taxa now considered to be in unrelated genera and families. Full descriptions supported by keys and b/w photographs, some uninformative, are
provided. Basidiospores are illustrated in line-drawings. A very useful
publication bringing together a large amount of information and little
used or even lost names. In French.

Singer, R. (1936). Studien zur Systematik der Basidiomyceten. Uber *Panus*
Fr. und verwandte Gattungen. *Beih. Bot. Zbl.* Abt. B. 56; 137-347. Pro-
vides keys to all the genera of white-spored agarics with excentric or later
stipe or with stipe lacking. A forerunner of the dissection of the genus
Pleurotus etc. In German.

Pilat, A. (1946). *Monographie especes europeenes du genre Lentinus Fr.: Atlas
des Champignons de l'Europe* 5. 46 pp., Prague. Full descriptions of
species supported by line-drawings and b/w photographs are offered. The
genus is taken in the widest sense but brings together a wide spectrum of
information. Supersedes the same author's publication in German in
Annls. Mycol. 34 (1936); 108-140.

Kuhner, R., Lamoure, D. and Fichet, M.L. (1962). *Geopetalum (Pleurotus)
longipes* (Boudier). Morphologie, Caryologie, Sexualite. *Bull. Soc. Mycol.
Fr.* 78; 135-154. Morphological and cytological data and discussion on
mating patterns in *Hohenbuehelia longipes* are given. Illustrated descriptive and experimental data are given. In French; summary in English.
See *Fungi Canadensis* No. 113. also Barron, G.L. and Dierkes, Y. (1977).
Can. J. Bot. 55; 3054-3062. A predatory state of *Nematoctonus* isolated

65

from a farmyard produced a bipolar mating *Hohenbuehelia*. Full experimental and illustrated descriptive data is given paralleling those of Kuhner, et al.

Kuhner, R., Lamoure, D. and Fichet, M.L. (1962). *Lentinus adhaerens* A. and S. ex Fries. Morphologie, Caryologie, Sexualite. *Bull. Soc. Mycol. Fr.* 78; 254-277. Morphological and cytological data and discussion on mating patterns in *Lentinus adhaerens* are given. Illustrated description and experimental data are given. In French, with English summary.

Miller, O.K. (1965). Three new species of lignicolous agarics in the Tricholomataceae. *Mycologia* 57; 933-945. Three new species of pleurotoid fungi are described and supported by illustrated descriptions; the species are each placed in *Lentinellus*, *Lentinus* and *Panus*.

Romagnesi, H. (1969). Sur les *Pleurotus* du groupe *ostreatus* (*Ostreomyces* Pilat). *Bull. Mycol. Soc. Fr.* 85; 305-314. Descriptions of members of the *Pleurotus ostreatus* complex are given and a discussion of the differences given. Special attention is paid to the presence of thick-walled hyphae. In French.

Miller, O.K. (1970). The genus *Panellus* in North America. *Mich. Bot.* 9; 17-30. A key to and illustrated descriptions of six species are given. Four species are included one being transferred to *Mycena*. Useful also to European workers.

Burdsall, H.H. and Miller, O.K. (1971). A re-evaluation of *Panellus* and *Dictyopanus* (Agaricales). *Nova Hedw., Beih.* 51; 79-91. A re-evaluation of these two genera is given; two poroid species are accepted in subgenus *Panellus*. A key is provided and illustrated descriptions offered.

Pegler, D.N. (1971). *Lentinus* Fries and related genera from Congo Kinshasa. *Bull. Jard. Bot. Nat. Belg.* 41; 273-281. The generic limits between *Lentinus*, *Pleurotus* and *Panus* are discussed. Several new combinations are made and two species described as new. In preparation for the Flore Champign. See pg. 84.

Kuhner, R. and Lamoure, D. (1972). Agaricales de la zone Alpine. Pleurotacees. *Botaniste* 55; 7-37. The new genus *Phaeotellus* is introduced. Eight species and varieties are described. Line-drawings are provided. In French.

Pegler, D.N. (1975). The classification of the genus *Lentinus* Fries (Basidiomycetes). *Kawaka* 3; 11-20. A revised outline classification of *Lentinus* is proposed based on the hyphal composition of the basidoiome. Four sections are recognized, one new and a second a new combination. The genera *Pleurotus* and the new genus *Lentinula* are defined, the latter contains the shiitake mushroom.

Singer, R. (1976). Marasmieae. *Flora Neotropica* 17. This contribution includes *Lachnella*, *Flagelloscypha* and *Amyloflagellula* (pg. 9-61). and *Gloiocephala*, *Physalacria*, *Epicnaphus*, *Manuripia*, *Humenogloea* and *Rimachia* (pg. 284-315). Keys, descriptions and discussion on both agaricioid and reduced members are given. Line-drawings are supplied and many new taxa proposed. Another important work bringing these reduced forms together; necessary for understanding relationships of temperate species referrable to this group.

Pegler, D.N. (1976). *Pleurotus* (Agaricales) in India, Nepal and Pakistan. *Kew*

Bull. 31; 501-510. A revision of the ten species recorded for the Indian subcontinent is given. A key is provided to the species and some descriptive data offered.

Hiland, K. (1976). The genera *Leptoglossum, Arrhenia, Phaeotellus* and *Cyphellostereum* in Norway and Svalbard. *Norw. J. Bot.* 23; 201-212. A report on the morphology, ecology and distribution in Norway and Svalbard of nine pleurotoid fungi in the above genera is given. Distribution maps and line-drawings of basidiomes and micro-structures are provided.

Singer, R. (1979). Key to species of *Cheimonophyllum. Sydowia* (1978) 31; 198-199. A key to two species plus one possibly distinct taxon is given. See Paxillaceae.

VOLVARIACEAE: see Pluteaceae above, pg. 40 ·

SCHIZOPHYLLACEAE

Schizophyllum
Cooke, W. Bridge (1961). The genus *Schizophyllum. Mycologia* 53; 575-599. *Phaeoschizophyllum* with one species, and *Schizophyllum* with five species are accepted and full descriptions offered. No line-drawings are given but a very good discussion on substrate-preferences and variation within *S. commune* is given. *P. lepieuri* is now placed in *Schizophyllum*. This paper supersedes the contribution by D.H. Linder, *Am. J. Bot.* 20; 552-564.

AURISCALPIACEAE and LENTINELLACEAE
See under hydnoid fungi for the first family and pleurctoid fungi for the latter. In the last case reference should be made to:

Pilat, A. (1946). *Monographie especes europeenes du genre Lentinus Fr. Atlas... de l'Europe* 5, see above, pg. 64.

CANTHARELLOID FUNGI: CANTHARELLACEAE
Also see Gomphoid fungi p. 68.
Characters. 1. Degree of development of ridges and veins. 2. Whether the pileus is perforate or not. 3. Spore-print colour. 4. Smell.

Donk, M.A. (1928). De geslachten *Cantharellus, Craterellus* en *Dictyolus* in Nederland. *Meded. Ned. Mycol. Vereen.* 16-17; 163-183. One of the first accounts to show the true relationship of *Cantharellus* and to *Craterellus* etc. and separation of the pleurotoid elements. In Dutch.

Smith, A.H. and Morse, E.E. (1947). The genus *Cantharellus* in the Western United States. *Mycologia* 39; 497-534. Full descriptions of the species recorded from the Western U.S. Supported by b/w photographs but no line-drawings. Includes elements now placed in *Gomphus.* see pg. 68.

Heinemann, P. (1958). Champignons recoltes au Congo Belge III. Cantharellineae. *Bull. Jard. Bot. Etat.* 28; 285-438. Descriptions and line-drawings of taxa recorded from the Congo; many new species proposed in preparation for the Flore Icon. see pg. 83.

Heinemann, P. (1959). *Flore Icon. Champignons Congo.* Fasc. 8. see pg. 83.
Heinemann, P. (1966). Cantharellineau du Katanga. *Bull. Jard. Bot. Etat.* 36; 335-352. A continuation of the author's earlier contributions on the Cantharelloid fungi of the Congo.
Corner, E.J.H. (1966). A monograph of Cantharelloid fungi. *Ann. Bot. Mem.* No. 2; 255 pp. A standard work covering several unrelated elements all brought together on their superficially similar hymenophore and blunt gill-margins. Full descriptions are given, supported by line-drawings. Several coloured plates are offered many depicting the new taxa proposed. Further notes see *Nova Hedw.* 18; 783-818 and Ibid 27 (1976); 325-342.
Smith, A.H. (1968). The Cantharellaceae of Michigan. *Mich. Bot.* 7; 143-183. Full descriptions accompanied by b/w photographs are given of all taxa recorded from Michigan; 22 species are described, five new. Useful to all workers, covering a greater area of temperate flora than indicated by title.
Perreau, J. (1970). Chanterelles et Craterelles. *Rev. Mycol.* 35; 280-286. A short descriptive key to the European species.
Petersen, R.H. (1971). Type studies in clavarioid fungi IV. Species from herbarium Fries at Uppaala with notes on cantharelloid species. *Friesia* 9; 369-388. A most valuable contribution redefining in modern terms some of Fries' collections.
Petersen, R. (1976). Notes on Cantharelloid fungi VII. Taxa described by C.H. Peck. *Mycologia* 68; 304-326. The most important article in a useful series on cantharelloid fungi by the author. The present paper covers a reassessment of Peck's types. See also *Persoonia* 5 (1969); 211-223.
Corner, E.J.H. (1976). Further notes on Cantharelloid fungi and *Thelephora*. *Nova Hedw.* 27; 325-342. An attempt to up-date the author's earlier publication.
Bigelow, H. (1978). The Cantharelloid fungi of New England and adjacent areas. *Mycologia* 50; 35-57. A summary of all the species so far recorded from the area indicated in the title, although of wider application. Full descriptions are given and supported by b/w photographs.

CLAVARIADELPHACEAE
Clavariadelphus
Wells, V.L. and Kempton, P. E. (1968). A preliminary study of *Clavariadelphus* in North America. *Mich. Bot.* 7; 35-57. Full descriptions and key are given to the species of the genus. Ten species are considered three new; b/w photographs and line-drawings are presented. See also Corner, E.J.H. (1950). A Monograph of *Clavaria* and allied genera. *Ann. Bot. Mem.* 1, with supplement in *Nova Hedw., Beih.* 33 (1970).

MERULIOID RELATIVES:
Heim, R. (1965). Les Meiorganes, Phylum reliant les bolets aux polypores. *Rev. Mycol.* 30; 307-339. The introduction of a new group of Basidiomycetes from New Caledonia showing relationships with Boletales. Full description, discussional and illustrative material offered. Latin diagnosis supplied in *Rev. Mycol.* 31 (1966); 157.

Corner, E.J.H. (1971). Merulioid fungi in Malaysia. *Gdns.' Bull. Singap.* 25; 355-381. *Meiorganum* and the Boletaceae are reassessed as is the *Merulius agathidis* and *Hygrophoropsis* and *Paxillus* relationship. *M. neocaledonicum* is re-described and illustrated with both line-drawings and coloured photograph. Eight species of *Merulius* and one of *Phaeophlebia* are also recognized and described.

CLAVARIOID RELATIVES

Coty, M.S. (1944). *Clavaria*, the species known from Oregon and the Pacific Northwest. *Oregon State Monograph: Studies in Botany* 4; 1-91. Full descriptions of all species with an initial attempt to divide them into groups now accepted as distinct, often unrelated genera. B/w photographs and line-drawings support the text.

Corner, E.J.H. (1950). A monograph of *Clavaria* and allied genera. *Ann. Bot., Mem.* 1-740 pp. (Suppl. in *Nova Hedw., Beih.* 33 (1970). A standard work covering practically all fungi at one time or another placed in *Clavaria* s. lato. Many unrelated elements are included. The genera covered include *Aphelaria, Araeocoryne, Ceratellopsis, Carpici, Chaetotyphula, Clavaria, Clavariachaete, Clavariadelphus, Clavulina, Clavulinopsis, Deflexula, Dimorphocystis, Hormomitra, Lachnocladium, Lentaria, Mucronella, Physalacria, Pistillaria, Pistillina, Pterula, Pterulicium, Ramaria, Ramariopsis, Sctytinopogon,* and *Typhula.* See also Thind, K. S. (1961). *Clavariaceae of India,* 197 pp., New Delhi; Coker, W.C. (1923). *The Clavarias of the United States and Canada,* Chapel Hill, 209 pp..

Perreau, J. (1969). Les Clavaires. *Rev. Mycol.* 33; 396-415. A descriptive key with a few supporting line-drawings. In French.

Petersen, R.H. (1967-1978). Type studies in the Clavarinaceae Series. I. The taxa described by C.H. Peck *Mycologia* 59 (1967); 767-802. II. *Nova Hedw.* 14 (1967); 407-414. III. The taxa described by J.B. Cleland. *Bull. Torrey Bot. Club* 96 (1963); 457-466. IV. Species from herbarium Fries...*Friesia* 9 (1971); 369-388. V. A few Australian taxa. *Mycotaxon* 7 (1978); 386-392. See pg. 67.

Petersen, R.H. (1967-1972). Notes on clavarioid fungi. Series. VII. Redefinition of the *Clavaria vernalis - C. mucida* complex. *Am. Midl. Nat.* 77. (1967); 205-221. VIII. *Clavaria pinicola* and *C. stillingeri. Bull. Torrey Bot. Club* 94 (1967); 417-422. XII. Miscellaneous notes on *Clavariadelphus* and a new segregate genus. *Mycologia* 64 (1972); 137-152. Three of the most important parts in this series from an agaricologists viewpoint. For part IV see pg. 67.

GOMPHACEAE AND RAMARIACEAE

GOMPHOID FUNGI: reference should also be made to Cantharelloid and Clavarioid fungi, pages 66 and 68 inc.

Corner, E.J.H. (1950). A monograph of *Clavaria* and allied genera. *Ann. Bot. Mem.* 1; 1-740: with supplement in *Nova Hedw., Beih.* 33 (1970); 1-299. see pg. 68.

Corner, E.J.H. (1966). A monograph of Cantharelloid fungi.*Ann. Bot. Mem.* 2; 1-255: with supplement in *Nova Hedw.* 27 (1976); 325-342. Includes full descriptions and illustrative data on members of the genus *Gomphus* along with members of the Cantharellaceae etc.

Petersen, R.H. (1968). Notes on Cantharelloid fungi I. *Gomphus* S.F. Gray and some clues to the origin of the ramarioid fungi. *J. Elisha Mitchell Scient. Soc.* 84; 373-381. One of a series of papers on both gomphoid and cantharelloid fungi.

Petersen, R.H. (1971). The genera *Gomphus* and *Gloeocantharellus* in N. America. *Nova. Hedw.* 21; 1-112. A good introduction with keys to subgenera, species etc. Seven species of *Gomphus* and one species of *Gloeocantharellus* are described; four new combinations are proposed and several new taxa. Line- drawings and coloured plates depicting eight taxa support the text. A companion text offering details of type-studies of twenty-four taxa is a very important contribution.

Marr, C.D. and Stuntz, D.E. (1973). *Ramaria* of western Washington. *Bibl. Mycol.* 38; 1-232. A monographic treatment with full supporting keys, b/w photographs of basidiomes and sporograms. Many new taxa and new combinations are proposed. Forty-seven species are described in full.

Petersen, R.H. (1974). Notes on Cantharelloid fungi V. Some fungi suspected of being Gomphoid. *J. Elisha Mitchell Scient. Soc.* 90; 53-54. *Phylloboletellus chloephorus* and *Linderomyces corneri* and considered.

Petersen, R.H. (1974-1976). Contributions towards a monograph of *Ramaria*. I. Some classic species redescribed. *Am. J. Bot.* 61 (1974); 739-748. Four species originally described by Schaeffer are redescribed. II. *J. Elisha Mitchell Scient. Soc.* 90 (1974); 66-68. III. *Ramaria sanguinea, R. formosa* and two new species from Europe. *Am. J. Bot.* 63 (1976); 309-316. The subject of mimic taxa is discussed.

Sharma, A.D. and Jandaik, C.L. (1978). The *Ramaria* in Himachal Pradesh. *Ind. J. Mushr.* 4; 5-7. Short descriptions are offered: see also Thind, K.S. (1961). *The Clavariaceae of India.* 1-197.

THELEPHOROID RELATIVES

Oberwinkler, F. (1976). Eine agaricoide Gattung der Thelephorales. *Sydowia* 28 (1975); 359-361. Recognition of a member of the Thelephorales amongst the agarics. The genus proposed is based on *Varrucispora* described by E. Horak in *Ber. Schweiz Bot. Ges.* 77 (1967); 362.

Corner, E. J. H. (1950). Clavaria and related genera . . . see pg. 68 for *Scytinopogon.*

Svrcek, M. (1958), Tomentelloideae Cechoslovakiae. *Sydowia* 14; 170-245. Covers forty species of *Tomentella*, and one each of *Tomentellina*, *Botryobasidium* and *Lindtneria*, six of *Pseudotomentella*, two of *Caldesiella* and five of *Tomentellastrum*. Keys to species and descriptions in Latin are given; introduction is in German.

Harrison, K.A. (1961). *Can. Dept. Agric.* Ottawa. Includes nine species of *Sarcodon*, three new (and as *Hydnum*) and fourteen species of *Hydnellum*, four new. Keys and coloured plates are given.

Harrison, R.A. (1964). New or little known North American Stipitate Hydnums. *Can. J. Bot.* 42; 1205-1234. Four new species of *Sarcodon* (as *Hydnum*) and ten of *Hydnellum* new described. B/w photographs are provided. Full descriptions of fresh collections of one further species in each genus are also provided.

Wakefield, E.M. (1966). Some extra-European species of *Tomentella*. *Trans. Brit. Mycol. Soc.* 49; 357-362. Eleven species, five of them new from Canada, Trinidad, Jamaica, Venezuela, South Africa and Pakistan are described.

Horak, E. (1967). Remarques critiques sur quelques champignons du Congo (Afrique). *Ber. Schweiz Bot. Ges.* 77; 262-266. *Lepiota verrucispora* is described - placed in *Horakia* by Oberwinkler see above.

Nikolajeva, T.L. (1968). The genus *Caldesiella* Sacc. in U.S.S.R. *Mikol. and Fitopatol;* 2; 198-202. A short account in Russian of this small genus.

Harrison, K.A. (1968). Studies on the Hydnums of Michigan. I. Genera *Phellodon, Bankera* and *Hydnellum. Mich. Bot.* 7; 212-264. Fifteen species of *Hydnellum* are included in this study; seventeen b/w photographs are provided.

Corner, E.J.H. (1968). A monograph of *Thelephora. Nova Hedw., Beih.* 27. A standard work with full descriptions, keys and line-drawings; includes six coloured plates. Forty-nine species, several new, are considered and an additional key is offered to species incertae sedis. (See supplement in *Nova Hedw.* 27 (1976); 325-342).

Wakefield, E.M. (1969). Tomentelloideae in the British Isles. *Trans. Brit. Mycol. Soc.* 53; 161-206. Forty species are described including two new species. In addition two species of *Caldesiella* and one of *Kneifiella* are described; line-drawings are provided. Supersedes Wakefield in *Trans. Brit. Mycol. Soc.* 5 (1917); 474-481, and is of wider application than title suggests.

Geesteranus, M.A. and Nannfeldt, J.A. (1969). The genus *Sarcodon* in Sweden in the light of recent investigations. *Svensk Bot. Tidskr.* 63; 401-440. Two new species are described. Nine species are dealt with in all; full descriptions and a key are provided. Line-drawings of sporograms are offered.

Geesteranus, M.A. (1971). *Hydnaceous fungi of the Eastern Old World*, London. Includes thirteen species of *Hydnellum* and nine species of *Sarcodon*. Three un-named species are included in the keys; coloured plates, line-drawings and descriptions are supplied. Important outside the area indicated by the title.

Larsen, M.J. (1974). A contribution to the taxonomy of the genus *Tomentella. Mycologia Memoir* 4. New York. Fourteen sections are described and keyed-out. Keys to the species within the sections are also given. Seventy-one species are described; in addition a useful list of nomina excludendae is given. Micrographs are provided. This monograph supersedes Larsen's publications in *Mycologia* 58; 597-613: 60; 670-679, in *Can. J. Bot.* 43; 1485-1510: 45; 1297-1307 etc. and *Tomentelloid fungi of North America. Tech. Publ.* 93, State Univ. College Forest. Syracuse Univ. Also see *Taxon.* 16; 510-511.

Geesteranus, M.A. (1975). *The terrestrial Hydnums of Europe*. London. Includes eight species of *Hydnellum*, two new and four species of *Sarcodon*. Keys, coloured plates, line-drawings and descriptions are supplied. Supersedes the same author's earlier works in *Fungus*, *Persoonia* and *Proc. K. ned. Akad. Wck.* Main text in German but a telescoped version in English is also provided. Covers other stipitate genera of '*Hydnum*'.

B. SPORE--STUDIES

D.N. Pegler and T.W. Young have carried out some rather interesting electron microscope studies on various agaric genera and correlated the characters found with the classic taxonomic features. The publications include micrographs and keys to species. - Basidiospore morphology in the Agaricales. *Nova Hedw., Beih.* 35. (1971).

Cantharellula	*Kew. Bull.* 28 (1973); 376.
Clitocybula	*Kew Bull.* 28 (1973); 326.
Clitopilus	*Kew Bull.* 30 (1975); 19 - 32.
Crepidotus	*Kew Bull.* 27 (1972); 311 - 323.
Galerina	*Kew Bull.* 27 (1972); 483 - 500. Ibid 30 (1975; 238).
Inocybe	*Kew Bull.* 26 (1972); 499-537.
Kuehneromyces	*Kew Bull.* 27 (1972); 498 - 499.
Leipsta	*Kew Bull.* 29 (1974); 659 - 667.
Leucopaxillus	*Kew Bull.* 28 (1973); 365 - 371.
Melanoleuca	*Kew Bull.* 28 (1973); 371-379.
Naucoria	*Kew Bull.* 30 (1975); 225 - 240.
Phaeogalera	*Kew Bull.* 30 (1975); 228.
Pleuroflammula	*Kew Bull.* 27 (1972); 322.
Poropoloma	*Kew Bull.* 28 (1973); 377.
Pseudoclitocybe	*Kew Bull.* 28 (1973); 377.
Pseudoomphvlina	*Kew Bull.* 28 (1973); 377.
Rhodocybe	*Kew Bull.* 30 (1975); 19 - 32.
Rhodotus	*Kew Bull.* 30 (1975); 19 - 32.
Ripartites	*Kew Bull.* 29 (1974); 659 - 667.
Simocybe	*Kew Bull.* 30 (1975); 228.
Entolomataceae.	World Pollen and Spore Flora, 7 (1978); Spore Form and Phylogeny of Entolomataceae *(Agaricales). Sydowia, Beih.* 8 (1979); 290 - 303.

III. REFERENCES ARRANGED IN GEOGRAPHICAL FRAMEWORK

A. GEOGRAPHIC DIVISIONS

The Geographical areas adopted herein follow the divisions of the herbarium at the Royal Botanic Garden, Edinburgh, Scotland.

1. Europe	Europe including U.S.S.R. as far east as the Urals, the Balkans.
2a. Orient	Turkey in Europe, Iran, Israel, Egypt, etc.

2b. N. Africa	Westwards from Egypt to the west coastal region.
3. N. Asia	Central and northern Asia, eastwards to Mongolia.
4. China	Including Korea, Formosa.
4a. Japan	
5. India	Including Pakistan, Burma, and Ceylon.
6. Malaya	Malaya penninsula, Thailand, Indo-china penninsula, Phillippines, Papua New Guinea.
7. Australia	Including Tasmania.
8. New Zealand	
9. Polynesia	Including Solomon Islands.
10. Tropical Africa	Central African States, Cape Verdi Is., former Portugese East Africa, Zimbabwe, Uganda, Zambia, Angola, etc..
11. Madagascar	Malagasy Rep., Maritius, and Seychelles.
12. South Africa	South Africa including Swaziland, Transvaal, Namibia and Tristan da Cunha.
13. North America	Canada, United States and Greenland.
14. Central America	Honduras, Costa Rica, Mexico.
15. West Indies	West Indies, Cuba and Bahamas.
16. E. Tropical S. America	Brazil, the Guianas and Paraguay.
17. W. Tropical S. America	Panama, Venezuela, Colombia, Ecuador, Peru, and Bolivia.
18. Temperate S. America	Chile, Argentina, Uruguay, Juan Fernandez.

B. REFERENCES

1. EUROPE: WEST AND NORTHERN EUROPE

Moser, M. (1978). In Gam's *Kleine Kryptogamenflora Mitteleuropa* IIb, 3rd edition 532 pp. An essential book, doubly useful as it is of pocket-size and can be taken into the field. Consists of descriptive keys supported by line-drawings of microscopic characters and habit-sketches. The keys are based primarily on field characters but genera and major sections are based on microscopic data; basidiospore-measurements are important. In German. The first edition in 1955 included useful addition of a key to Gasteromycetes omitted from the second (1967) and third editions, although the latter includes secotiaceous elements. The third edition is considerably enlarged.

British Isles
Rea, C. (1922). *British Basidiomycetae*, Cambridge, 799 pp. Full descriptions of all then known British basidiomycetes including agarics. Out-of-date

but still exceedingly useful. Reprinted as *Bibl. Myco.* 15 (1968), J. Cramer.

Rea, C. (1927 and 1932). Appendices to British Basidiomycetae. - I. Additions and corrections. *Trans. Brit. Mycol. Soc.* 12; 205-230. II. Ibid 17; 35-50, including one coloured plate.

Pearson, A.A. (1938-1951). Agarics: New records and observations. I. *Trans Brit. Mycol. Soc.* 22; 27-46 with coloured plate. Two new species. II. Ibid 26; 36-49. Several new combinations proposed. III. Ibid 29; 191-210. Three new species and several infra-specific taxa proposed; three b/w plates and two coloured plates presented. IV. Ibid 32; 258-271. One new species described; line-drawings and b/w photographs given. V. Ibid 35; 97-122. Seven new species are described; one coloured plate offered. Full descriptions of new and critical species are presented. A very useful series paper.

Dennis, R.W.G. and Pearson, A.A. (1948). Revised list of British Agarics and Boleti: see Dennis, Orton and Hora below.

Orton, P.D. (1957). Notes on British Agarics. *Trans. Brit. Mycol. Soc.* 40; 263-267: see below for list of parts in this series.

Dennis, R.W.G., Orton, P.D. and Hora, F.B. (1960). New Check List of British Agarics and Boleti. *Trans. Brit. Mycol. Soc.* 43. Suppl. including Part III. Notes on genera and species in list. *Trans. Brit. Mycol. Soc.* 43; 159-439. (P.D. Orton). Includes keys to *Crepidotus, Flammulaster, Hygrophorus, Naucoria, Nolanea* and *Pleutus*; also incorporates Part IV. Validations, new species, and critical notes (F.B. Hora). Ibid 43; 440-459. Full descriptions and supporting line-drawings.

Supersedes Revised List . . . (A.A. Pearson and R.W.G. Dennis): *Trans. Brit. Mycol. Soc.* 31 (1948); 145-190. Some of the fungi analyzed for the Revised List are considered in Dennis, R.W.G. (1948). Some little known British species of Agaricaceae I. Leucosporae and Rhodosporae. *Trans. Brit. Mycol. Soc.* 31; 91-209. Check List reprinted as *Bibl. Mycol.* 42. (1974). J. Cramer.

Orton, P.D. Series- Notes on British Agarics, with line-drawings and full descriptions.

1. *Trans. Brit. Mycol. Soc.* 40 (1957); 263-267. *Coprinus* species only.

2. *Notes R. Bot. Gdn. Edinb.* 26 (1964); 43-65. Notes on many new and critical species.

3. Ibid. 29 (1969); 72-127. Notes on genera in British flora and key to British *Psilocybe* supp.

4. Ibid. 32 (1972); 135-150. *Coprinus* species considered.

5. *Kew Bull.* 31 (1976); 709-721. Notes on new and critical species.

6. *Notes Roy. Bot. Gdn. Edinb.* 35 (1976); 147-154 with appendix by Roy Watling. Covers various species, many new, of *Coprinus*; in preparation for The British Flora. Also includes analysis of British blue-green *Stropharia* spp.

7. *Notes R. Bot. Gdn. Edinb.* (1980); in press. Notes on critical species, several new, are given; includes key to *Deroloma*.

Henderson, D.M., Orton, P.D. and Watling, R. (1968-   ). *British Fungus Flora: Agarics* and *Boleti*. Her Maj. Stat. Office, Edinburgh. Includes keys, full descriptions and line-drawings. An attempt in parts to cover the

74

entire agaric flora. Introduction. Keys to families and genera of agarics
found in the British Isles; glossary, methods of collecting, etc.
 Part I. Boletaceae, Gomphidiaceae and Paxillaceae (by Roy Watling).
 Part II. Coprinaceae: *Coprinus* (by P.D. Orton and Roy Watling).
 Part III. Bolbitiaceae (by Roy Watling - in press).
 Part IV. Russulaceae: *Lactarius* (by R. Rayner - in preparation).
Pearson, A.A. et. al. (1978). Naturalist Keys to Agaric Genera; inspired by
 A.A. Pearson and including his paper 'Diagnostic Characters in the
 Agarics'. Introduction and commentary by Roy Watling. Includes keys by
 P.D. Orton, R.W.G. Dennis and F.B. Hora.
Watling, R. (1967-1974). Notes on some British Agarics. Series. *Notes R. Bot.
 Gdn. Edinb.* I. Covers records of *Coprinus, Hygrophorus* and *Russula* spp.
 28; 39-56. II. Three *Coprinus* spp. are described. *31;* 359-363. III. Four
 Coprinus spp. are described. *32*; 127-134. IV. One species of *Boletus* and
 one of *Squamanita* are described. *33*; 325-332.

France

Konrad, P. and Maublane, A. (1924-1937). *Icones Selectae Fungorum,* Six
 volumes, Paris; also *Les Agaricales. Encycl. Mycol.* 14 and 20, Paris, 469
 pp. and 202 pp. respectively. Very useful compilations covering W. Continental Europe and in part N. Africa.
Josserand, M. (1933-1974). Series. Notes critiques sur quelques champignons
 de la region Lyonnaise. In French. see *Bull. Mycol. Soc. Fr.* 49 (1933);
 340-375: Ibid 53 (1937); 173-230: Ibid 59 (1943); 5-33: Ibid 64 (1948); 5-32:
 Ibid 71 (1955); 65-125: Ibid 75 (1959); 359-404: Ibid 81 (1965); 517-565:
 Ibid 90 (1974); 231-263. A very useful series of papers covering various
 agaric genera with full illustrated descriptions. Several new species and
 new names proposed.
Kuhner, R. and Romagnes, H. (1953). *Flore analytique des Champignons
 Superieurs,* Paris, 556 pp. The standard European agaric Flora in
 French; essential to all students in the field. Reprinted by Maisson in
 1974. Supplements: - Complements a la Flore analytique. In French.
 I. Entolomataceae - all entries under *Rhodophyllus.* Especes nouvelles ou
 critiques de *Rhodophyllus. Rev. Mycol.* 19 (1954); 1-46. Twenty-three
 Latin diagnoses are supplied; only with discussion on twenty species
 and varieties. Supported by line-drawings. Especes ..., Descriptions
 et remarques de H. Romagnesi. *Rev. Mycol.* 20 (1955); 197-230. Thirteen species and varieties are discussed. Supported by line-drawings
 and one plate.
 II. Russulaceae
 Especes nouvelles ou critiques de *Lactarius. Bull. Soc. Mycol. Fr.* 69
 (1954); 361-388. Eight Latin diagnoses supplied and ten species.
 III. Tricholomataceae
 Especes nouvelles, critiques ou rares de Pleurotacees, Marasmiacees
 et Tricholomacees. *Bull. Soc. Nat. Oyonnax* 8 (1954); 73-131. Fifteen
 Latin diagnoses are supplied. Thirty-four descriptions are given with
 some line-drawings.
 IV. Cortinariaceae
 Especes nouvelles ou critique de *Cortinarius. Bull. Soc. Linn. Lyon*

24 (1955); 39-54 (by R. Kuhner). Latin descriptions are given for ten taxa; fifteen taxa are discussed and supported by line-drawings.

V. *Inocybe* leiospores cystidies. *Bull Soc. Nat. Oyonnax* 9 (1956); 3-95, (by R. Kuhner). Twenty-four Latin diagnoses offered. Forty detailed descriptions are given supported by useful illustrations.

VI. *Inocybe* goniospores et *Inocybe* acystidies. *Bull. Soc. Mycol. Fr.* 71 (1956) 169-201 (by R. Kuhner). Nine Latin diagnoses to new taxa are given. Fifteen taxa are described with supporting line-drawings.

VII. Especes nouvelles, critiques ou rare de Naucoriacees, Coprinacees et Lepiotacees. *Suppl. Bull. Soc. Nat Oyonnax* 10-11 (1957); 4-57 and 65-91. Ten Latin diagnoses are offered; twenty detailed descriptions are given by Kuhner with line-drawings covering species now placed in *Simocybe, Flammulaster, Hebeloma, Galerina, Phaeocollybia, Gymnopilus* and *Pholiota*. Ten descriptions of species in *Tubaria, Flammulaster, Gymnopilus* and *Pholiota* are given by Romagnesi.

Coprinaceae

Especes nouvelles, critiques ou rare de Naucoriacees, Coprinacees et Lepiotacees. *Suppl. Bull. Soc. Nat. Oyonnax* 10-11 (1957); 3 and 57-65. VII--Four Latin descriptions are given, one in *Psathyrella* and three in *Coprinus*. Five descriptions are given by Kuhner.

Lepiotaceae

Especes nouvelles, critiques ou rare de Naucoriacees, Coprinacees et Lepiotacees. *Suppl. Bull. Soc. Nat. Oyonnax* 10-11 (1957); 91-94. VII-Two descriptions supplementary to the 'Flore' are given by H. Romagnesi.

VIII. Pluteaceae

Especes nouvelles, critiques ou rare de Volvariacees. *Bull. Soc. Mycol. Fr.* 72 (1956); 181-249. Seven Latin diagnoses are given. Twelve descriptions of *Pluteus* are given by Kuhner and nine by Romagnesi. Four descriptions of species of *Volvariella* are given in addition.

Complements a la Flore ... have been brought together in *Bibl. Mycol.* 56 (1977).

Bon, M. (1970-1977). Macromycetes de la zone Maritime Picarde. Series. I. Flore Heliophile. *Bull. Soc. Mycol. Fr.* 86; 79-213. Coloured plates and line-drawings are included. II. *Docums. Mycol.* 5 (19): 1975; 21-36. III. Le genre *Hygrocybe*. Especes critiques rare ou nouvelles et rivision du genre. *Bull. Soc. Mycol. Fr.* 93; 201-232. Key to sections of *Hygrocybe* and to the taxonomic 'splits'. IV. *Docums, Mycol.* 7; 63-80. Key to *Tephrocybe* is presented. In French.

Bon, M. (1970-1973). Macromycetes du Nord de la France. Series I. Cantharellaceae Hygrophoracees. *Bull. Soc. Bot. Nord Fr.* 23 (1970); IIa. Boletacees and *Russula* spp. *Rev. Mycol.* 35 (1970); 229-257. IIb. *Russula* (cont'd.) and *Lactarius*. *Rev. Mycol.* 38 (1973); 185-206. III. *Pholiota* spp. formerly placed in *Flammula*; includes a key. *Bull. Soc. Bot. Nord Fr.* 24 (1971); 43-60.

Maublanc, A. and Viennot-Bourgin, G. (1971). *Champignons comestibles* et *veneneux* Paris, Vol. 1 text, 305 pp.; Vol. 2. Atlas, 226 plates with accompanying text and descriptions. 249 pp. Very useful illustrations although somewhat stylized. Still a very popular book now enjoying its sixth edition. In French.

Bon, M. and Chevassut, G. (1973). Agaricales de la region Languedoc - Cevennes. I. *Docums. Mycol.* 3 (9) 1-50. II. *Docums. Mycol.* 3 (11); 1-29. III. *Docums. Mycol.* 4 (15); 1-35. Illustrated descriptions are provided.

Bon, M. (1975). Agaricales de la Cote Atlantique Francaise. *Docums. Mycol.* 4 (17); 1-40. Full descriptive data accompanies Bon's articles; several new taxa are proposed and line-drawings of microscopic characters and habit-sketches are included. All in French but of wider implication than the various titles would appear to indicate.

Romagnesi, H. (1976-   ). Series. Quelques especes meconnues ou nouvelles de Macromycetes. In French. Title often changes to . . . especes rares ou nouvelles 1111 I. Coprinacees, *Bull, Soc. Mycol. Fr.* 92 (1976); 189-206. Four species of *Coprinus* and *Psathyrella* are described and a key to sect. *Micacei* of *Coprinus* is given. II. Ibid 92 (1976); 299-209. Rhodophyllees. Three taxa are considered; a key to sect. *Lampropodes* is given III*. Trav. dedies a G. Viennot-Bourgin; 337-344. *Cortinarius*: Three new species of *Cortinarius* are described. III*. *Sydowia, Beih.* 8; 349-365. *Inocybe:* Eight taxa, six new are considered. V. *Bull. Soc. Mycol. Fr.* 94 91978); 73-85. Agarics leucospores. Four new taxa are described and two new combinations are made. VI. Ibid 94 (1978); 97-108. Four taxa, two new, are described in *Armillariella, Mycena, Rhodocybe* and *Entoloma*. VII. Ibid (1979); 37-378. Agarics rhodospores (Volvariacees). Three new species (one in *Volvaria* and two in *Pluteus*) are described. The series is supported by line-drawings of microscopic characters in habitat sketches are given. In French. See also Romagnesi, H. (1974). *Bull. Soc. Mycol. Fr.* 90; 161-169. Quelques especes et varietes mecconnus d'Agaricales. Five taxa, one new all but one in the Tricholomataceae. *Note parts are both numbered III.

Alpine studies

Kuhner, R. Agaricales de la zone alpine. See under appropriate genus-heading for full references. Also see J. Favre under entry heading Switzerland.

1970. Introduction with Mlle. D. Lamoure. *Bull. Soc. Mycol. Fr.*
86; 875-880.

1971. *Rhodocybe.*

1971. Rhodophyllacees caracteres generaux et grandes lignes de la classification.

1971. *Rhodocybe borealis* (D. Lamoure)

1972. *Amanita* and *Limacella*

1972. Pleurotacees with D. Lamoure

1972. *Clitocybe* (D. Lamoure)

1972. *Galerina*

1972. *Galerina* (conclusion) and *Phaeogalera*

1973. *Mycena ochrogaleata* (D. Lamoure)

1974. *Agaricus*

1974. *Omphalina* Part 1 with D. Lamoure

1975. *Omphalina* Part 2 with D. Lamoure

1975. *Lactarius*

1975. *Russula*
1976. Lepistees (*Ripartites* and *Lepista*)
1976. *Hygrocybe* Part 1
1977. *Hygrocybe* Part 2
1977. *Psilocybe chionophila* (D. Lamoure)
1977. *Cortinarius* subgenus *Telamonia* Part 1 D. Lamoure
1977. Hygrophoracees; generalites and *Camarophyllus*
1977. Rhodophyllacees
1978. *Melanoleuca*
1978. *Cortinarius* subgenus *Telamonia* Part 2 D. Lamoure

Switzerland

Favre, J. (1948). *Materiaux pour la Flore Cryptogamique Suisse*. Vol. 10 (3); 1-228. Les associations fongiques des hauts-marais Jurassiens et de quelques regions voisines. Berne. Six coloured plates and sixty-seven text figures; very full descriptions.

Favre, J. (1955). Les Champignons superieurs de la zone alpine du Parc National Suisse. *Publi. Comm. Soc. Helvetique des Sciences Naturelles pour les etudes scientifiques du Parc National*, 5: 37; 1-212. Liestal. Eight coloured plates, five b/w plates and one hundred and forty-five text figures; full descriptions.

Favre, J. (1960). Catalogue descriptif des Champignons superieurs de la zone subalpine du Parc National Suisse, Ibid 6: 52; 1-610. Eight coloured plates and one hundred and four text figures. A trio of papers essential to all those working on alpine and northern fungi and an example of good alpha taxonomy applied to agarics. Probably of wide application in northern and Andean zones.

OTHER EUROPEAN COUNTRIES IN ALPHABETICAL ORDER

Atlantic Islands

Moeller, F.H. (1945). *Fungi of the Faeros*. Copenhagen, 294 pp. Good descriptions accompanied by text-figures and three coloured plates.

Moeller, F.H. (1958). *Fungi of the Faeroes*, appendix, 285 pp. As above but with one additional coloured plate of non-agarics.

Czechoslovakia

Velenovsky, J. (1920-22). *Ceske Houby* 950 pp. Prague, Prepared in five parts. Many score new species described in Czech. Discussion in Czech.

Vesely, R. (1938). *Ceske Houby* I. Lupenate, Prague. *Ceskoslovenske Houby* II. Chlorosovite a dalsi stopkovytruse, vreckate, Prague.

Melzer, V. (1945). Atlas holubinek. Ceske. holubinky (Russulae Bohemiae) *Archiv. pro prirodovedecky Vyzkum Cech* 17 (109); 1-126.

Velenovsky, J. (1947). *Novitates Mycologicae Novissimae*. Prague *(Opera Botanica Cechica* 4; 3-158). Introduction and preface in English. Many hundred new species, over half agarics, are described. One plate of line-drawings supplied. In Latin with discussion in German.

Vacek, V. (1948). The Bohemian and Moravian species of the genus *Pluteus*. *Studia Bot. Cechoslovaca* 9; 30-48.

Pilat, A. (1952 and 1959). *Nase Houby,* Vol. I and II, Prague.

Pilat, A. (1953). Hymenomycetes novi vel minus cogniti, Cechoslovakiae II. *Sbornik Narodnitio musea v* Praze 9; B2, Bot. No. I.

Zvara, M. and Zvarova, M. (1966). *Zbierame huby.* Bratislava.

Prihoda, A. (1973). *Hubariv rok,* includes 214 coloured plates.

Denmark

Lange, J. (1914-1940). *Flora Agaricina Danica* I-V. Copenhagen. (first produced as separate parts in *Dansk. Bot. Arkiv.* 1-9. Includes five hundred fine coloured plates supported by descriptions, some unfortunately rather sparse and based largely only of macroscopic characters. In English. Some plates although often of inferior reproductive quality are found in Colins' popular guide by M. Lange (and F.B. Hora for use in the British Isles). see pg. 100.

Finland

Ulvinen, T. (ed.) (1976). *Suursieniopas Suomen Sieniseura.* Helsinki, 359 pp. Descriptive keys to the majority of Finnish larger fungi including some Ascomycetes; with maps and line-drawings. In Finnish.

Germany

Ricken, A. (1915). *Die Blatterpilze,* Part I (Text) and Part 2 (coloured plates) Leipzigs 48 pp. Full descriptions with very useful information on microscopic characters. Although the use of some of the epithets is uniquely that of Ricken the author's descriptions are so good that these species can be recognized. A classic work in German.

Iceland

Christiansen, M.P. (1941). Studies in the larger fungi of Iceland. *Botany Icel.* III, 2; 191-226. List of agarics with limited descriptive data.

Hallgrimsson, H. (1973). Islenzker hathsvepper III. Lepiotaceae, Gomphidiaceae, Paxillaceae and Crepidotaceae. *Acta Bot. Isl.* 2; 29-55. List of species with limited descriptive data.

Italy

Bresadola, G. (1881). *Fungi Tridentini.* Tridenti. Reprinted by I. Bresadola for Edagricole. Coloured plates of individual species will full descriptive data.

Saccardo, P.A. (1915-1916). *Flora Italica Cyrptograma* 1. Fungi hymeniales. Part I 1955. Part 2. (1916). *Soc. Bot. Italiana.* An exceedingly useful compilation with descriptions. No illustrations of use in identical.

Netherlands

Huijsman, H.S.C. (1955). Observations on agarics. *Fungus* 25; 18-43. Various agarics especially in *Inocybe* and *Leucoagaricus.* B/w photographs and line-drawings are supplied to support the descriptive data.

Spain

Heim, R. (1934). Fungi Iberica: Obs. sur la flore mycologique Catalane.

Treballs del Mus. Ciencies Nat. Barcelona 15; = 146. 1-146. Line-drawings and four coloured plates with accompanying text. In French.

Maire, R. (1933). Fungi Catalaunici cont. a l'etude de la Flore de la Catalogne. *Mus. barcin. Scient. nat. op.* 15; -1-120. One coloured plate accompanying text. In French.

Singer, R. (1947). Champignons de la Catalogne. *Collnea. Bot. Barcinone* 1; 199-246. A useful contribution from this very important and rich area of Europe. In French.

Maire, R. (1937). Second 'list'. *Pub. Inst. Botanic de Barcelona* 3; 1-128. Line-drawings accompanying text; in French.

Spitsbergen

Kobayasi, Y., Tubaki, K. and Soneda, M. (1968). Enumeration of the Higher Fungi... Spitsbergen. *Bull. Nat. Sci. Mus. Tokyo* 11; 33-76. Twenty-four species of agaric and allied fungi are listed with short descriptions and some line-drawings. Some collections are only identified to genus. A very useful contribution which up-dates Dobbs' study in *Journ. Bot.* (1942); 94-102 and enumerates the species in modern terms.

MEDITERRANEAN AREAS (2A AND 2B)

2A ISRAEL

Binyamini, N. (1945). *Fleshy fungi of Israel* (Agaricales). Tel Aviv, 225 pp. An illustrated guide to two hundred and seventy local species with ninety-six coloured photographs and sixty-four b/w illustrations. A useful guide; in Israeli.

Binyamini, N. (1974). Fleshy fungi of North and Central Israel I. *Israel J. Bot.* 23; 237-251. Thirty-eight taxa of fleshy fungi growing in association with oak and pine are described. All are new records for Israel.

2B NORTH AFRICA
Morocco

Malenson, G. and Bertault, R. (1970). *Flore des Champignons superieurs du Maroc.* I. Fac. Sciences Rabat. Ouvrage Public avec le concours du centre National de la Recherche scientifique. Covers Amanitoideae, Volvariodeae, Lepiotoideae, Coprinoideae, Naucoriodeae (Bolbitieae, Geophileae and Cortinarieae) and Rhodophylloideae.

Malenson, G. and Bertault, R. (1975). Ibid II. This volume covers Tricholomoideae (Orcelleae, Lyophylleae and Tricholomeae). Pleurotaceae, Hygrophoraceae, Cantharellaceae and Gomphaceae. Unfortunately because of lack of finance the series will not be continued except as parts appearing in journals. See *Sydowia Beih.* 8; 258-267. Four species are considered, two new.

U.S.S.R. (1 AND 3)

Kalamees, K. (1971-1972). Eesti seente Maaraja, Tartu; in Estonia, Keys to agarics with small line-drawings. I. Cantharellaceae, Pleurotoid fungi, Hygrophoraceae, Tricholomataceae, Agaricaceae, Entolomataceae, Pax-

illaceae and Gomphidaceae, boletes and jelly-fungi; 229 pp. II. Coprinaceae, Bolbitiaceae, Stropharicaceae, Cortinariaceae and Crepidotaceae;
132 pp. A full agaric flora with descriptive keys.

Leisner, R. (1973). *Eesti Pilvikid.* Tallin 40 pp. An account of Estonian
Russula spp. with twenty-four coloured plates of watercolours. In
Estonian.

Kalaamees, K. and Lasting (1974). (Strobilomycetaceae, Gyrodontaceae and
Boletaceae), Tallin, 46 pp. An account with thirty-two coloured plates of
boletes recorded for Estonia in Estonian with summary in English.

Wasser, S. (1977). (Boletales, Agaricales, Russulales and Aphyllophorales), Kiev. Numerous line-drawings and b/w photographs. Text in
Ukrainian; 356 pp.

Kalamees, K. (1979). *Eesti Riisikad.* Tallin, 62 pp. An account of Estonian
Lactarius spp. with thirty-two coloured plates of watercolours. In Estonian. A key is provided. Summary in English.

a) Caucasus

Singer, R. (1929). Pilze aus dem kaukasus 1. *Beih. Bot.* Zbl. 46 (1929); 71-113:
II. Ibid 48 (1931); 513-542. A very useful contribution to the agaric flora of
this part of Europe and the first major publication from Rolf Singer after
his work on *Russula*. Indicates the direction in which Singer's studies
were going. In German.

4. HONG KONG

Griffiths, D.A. (1977). Fungi of Hong Kong. Hong Kong; 129 pp. Lavishly
illustrated book of popular appeal within limited text and little or no
descriptive data. Ninety-nine fungi are illustrated the first sixty-four
being agarics and boleti some incorrectly identified.

JAPAN (1-4A)

Imai, S. (1933-1935). Studies on the Agaricaceae of Japan. I. Volvate agarics
in Hokkaido. *Bot. Mag., Tokyo* 47 (1933); 423-432. II. Lactarius in Hokkaido. Ibid 49 (1935); 603-610. see below.

Imai, S. (1939-1941). Studia Agaricacearum Japonicarum. I. *Bot. Mag. Tokyo*
53 (1939); 392-399. II. Ibid 55 (1941); 444-451. III. Ibid 55 (1941); 514-520.
Some lists of fungi are offered, some descriptions and some discussional
material. Continuation of Studies on Agaricaceae.

Imai, S. (1938). Studies on the Agaricaceae of Hokkaido. *J. Fac. Agric. Hokkaido (Imp.) Univ.* 43; 1-378. A substantial publication in which many
new combinations are suggested and old traditional genera more pre-
cisely delimited with type species cited. Five plates support the article.

Hongo, T. (1955-1974). Notes on Japanese larger fungi. *J. Jap. Bot.* 30-49.
Descriptions of various agaric taxa supported by line-drawings. In English; Japanese summary.

Imazeki, R. and Hongo, T. (1957). *Coloured illustrations of Fungi of Japan.*
Osaka, 166 pp. Beautiful coloured illustrations with accompanying line-
drawings and descriptive text. In Japanese. Reprinted and up-dated in
1969 and in 1978 as one of a pair of books; see below.

Hongo, T. (1959-1960). The Agaricales of Japan. *Mem. Shiga Univ.* 9-10. Descriptions of various agaric taxa supported by line-drawings. In English; Japanese summary.

Hongo, T. (1962-1977). Notulae Mycologicae 1-13. *Mem. Shiga Univ.* 12-27. New taxa in various agaric genera supported by line-drawings. In English; Japanese summary.

Hongo, T. (1977-1978). Higher Fungi of the Bonin Islands I. *Mem. Natn. Sci. Mus., Tokyo* 10; 31-41. Higher fungi... II. *Repr. Tottori Mycol. Inst.* 16; 59-65. Same format as earlier contributions by the author.

Imazeki, R. and Hongo, T. (1978). *Coloured Illustrations of Fungi of Japan.* Osaka. Vol. II: 224 pp. In Japanese. One of a pair; the first volume being issued earlier. Same format with beautiful coloured plates accompanying text.

See also Hongo, T. Other series by this same author with or without co-authors are 'Materials for the Fungus of flora of Japan', 1-12 (1960-1973). *Trans. Mycol. Soc. Japan* and 'On some Agarics of Japan ,' *Mem. Shiga Univ.*

5. INDIA AND SRI LANKA

Sathe has listed many agarics from the S. West region of India in recent studies but has offered as yet no descriptive data. Also see Pegler under Pleurotaceae and Lepiotaceae and Petch under *Marasmius* (Tricholomataceae).

Natrajan, J. (1975-1979). South Indian Agaricales. I. Four species of *Termitomyces. Kavaka* 3; 63-66. II. *Gymnopilus tropicus* and *Galerina truncata. Mycologia* 69 (1977); 185-189. One species each in *Amanita, Leucocoprinus* and *Termitomyces* and two species in *Laccaria* are supplied with descriptions and line-drawings. *Kavaka* 5; 63-66. IV. *Psathyrella. Mycologia* 70 (1978); 1259-1261. V. *Termitomyces heimii. Mycologia* 71 (1979); 853-855. Full descriptive data supported by line-drawings.

Watling, R. (1980 in press). A new hypogeous *Cortinarius. Notes R. Bot. Gdn. Edinb.* A new species is illustrated and described from Kashmir. A key to all the hypogeous taxa so far known is given.

Watling, R. and Gregory, N. (1980 in press). Larger fungi of Kashmire. *Notes R. Bot. Gdn. Edinb.* Full records of agarics and related fungi collected in Kashmir during autumn 1977 by the senior author are given. Many collections are described and several line-drawings are provided.

6. S. E. ASIA

New Guinea and Solomon Islands

Hongo, T. (1974-1976). Agarics from Papua - New Guinea 1 - 3. *Rept. Tottori Mycol. Inst.* Japan 1-14. Several new species are described. Full descriptions are offered and supported by line-drawings. In English.

Hongo, T. (1973). Mycological Reports from New Guinea and the Solomon Islands. Reports. 14, 15, and 21.

14. *Rept. Tottori Mycol. Inst., Japan* 10; 341-356. Twenty species described seven of which are agarics. Line-drawings are provided.

15. *Rept. Tottori Mycol. Inst., Japan* 10; 357-364. Seven taxa, two new are

fully described and supported by line-drawings.

21. Enumeration of the Hygrophoraceae, Boletaceae and Strobilomyceta-ceae. *Bull. Nat. Sci. Mus., Tokyo* 16; 543-557. Fourteen boletes and allies, three new and two unidentified, are described and seven members of the Hygrophoraceae, one new and two unidentified, are detailed; line-drawings are offered. In English.

Horak, E. (1976). *Boletellus* and *Porphyrellus* in Papua - New Guinea. *Kew Bull.* 31; 645-652. see pg.

Horak. E. (1979). Three new genera of Agaricales from Papua New Guinea. *Sydowia, Beih.* 8; 202-208. One genus each in Cortinariaceae, Agarica-ceae and Lepiotaceae (?) are provided with illustrated descriptions.

Thailand

Hein, R. (1962). Contribution a la flore mycologique de la Thailande. *Rev. Mycol.* 27; 116-158. Illustrated descriptions of larger fungi are offered; one coloured, three b/w and line-drawings support the text. In French.

7. AUSTRALIA AND TASMANIA

Cleland, J. W. (alone or in collaboration with E. Cheel)(1918-1931). Australian Fungi: Notes and descriptions. Series *Trans, Proc. R. Soc. S. Aust.* 42 (1918); 88-138. II. Sclerotia forming polypores of Australia. Ibid 43 (1919); 11-22. III. Ibid 43 (1919); 262-315. IV. Ibid 47 (1923);58-128. V. Ibid 48 (1924); 236-252. VI. Ibid 51 (1927); 298-306. VII. Ibid 52 (1928), 217-222. VIII. Ibid 55 (1931); 152-160. IX. Ibid 59 (1935); 219-300. The last three parts deal solely with non-agarics. See also Hymenomycetes of New South Wales in *Agri. Gaz. New S. Wales* 25; 507-515; 885-888; 1045-1049: Ibid 26; 325-333 and ibid 27 (1916); 97-106. Apparently these are the only articles which appeared. Descriptions of *Amanita* (including *Amanitop-sis*), *Lepiota* and of *Armillaria* are offered.

Cleland, J. (1934). *Toadstools and Mushrooms and other larger Fungi of South Australia.* Adelaide, 362 pp.. A standard two volume text on agarics of s. Australia but used throughout the island continent to identify agarics; relied on too heavily by agaricologists as an identification manual and by European and N. American mycologists as a measure of the flora. Too many European epithets are adopted. Reprinted in 1976 by S. Australian Govt., with introduction by P. Talbot. Very useful if the restrictions are appreciated. This work was based on the experience gained by the author in South Eastern Australia some of the results being published in various papers.

Pegler, D.N. (1965). Studies on Australian Agaricales. *Aust. J. bot.* 13; 323-356. A most valuable contribution reporting on an analysis of all those types of Australian agarics at Kew. Microscopic data is offered which has never previously been published.

8. NEW ZEALAND

Stevenson, G. (1961-1962). The agarics of New Zealand. I. Boletaceae and Strobilomycetaceae. *Kew Bull.* 15; 381-385. Nine species, four new are described and accompanied by one line-drawing and one coloured plate. II. Ibid 15; 65-74. Amaniteae and Pluteae. Seventeen species, ten new (three in *Amanita*, six in *Pluteus* and one in *Limacella*) are described and supported by two plates of line-drawings and three coloured plates. III. *Rhodophyllaceae*. Ibid 16; 227-237. Twenty-four species, nineteen new, are described and supported by two coloured plates and two platees of line-drawings. IV. Hygrophoraceae. Ibid 16; 373-384. Twenty-five species, seventeen new, are described and supported by three coloured and one black and white plate.

Taylor, M. (1970). *Mushrooms and Toadstools in New Zealand.* Wellilngton 32 pp. An introduction to New Zealand fungi with short descriptions and coloured illustrations. Although intended to be a popular account it is very useful to the visiting professional agaricologist.

Horak, E. (1971). Contributions to the knowledge to the Agaricales of New Zealand. *N. Z. J. Bot.* 9; 463-493; see also *N. Z. J. Bot.* 9; 403-462.

Horak, E. (1973-1979). Fungi Agaricini Nova Zealandiae I-V. Issued as a separate publication *Nova Hedw., Beih.*, 43. See under separate generic headings; also Part VI. *Inocybe* and *Astrosporina*. VII. *Rhodocybe* see *N. Z. J. Bot.* 15; 713-747; Ibid 17; 275-282. I. *Entoloma* and related genera. II. *Thaxterogaster*, III.*Hygrophorus* and related genera. IV. *Phaeocollybia* V. *Cuphocybe*.

10.TROPICAL AFRICA AND ASSOCIATED AREAS

Flore Illustrate Champignons d'Afrique du Congo; replacing in 1972 'Flore Iconographique des Champignons du Congo'. See under each individual generic heading; species usually supplied with Latin diagnoses in *Bull. Jard. Bot. Etat.* Beautifully produced and lavishly illustrated with coloured plates. In French.

Fasc. 1. 1935. *Amanita* and *Volvari (ell) a*; M. Beeli
Fasc. 2. 1936. *Lepiota* and *Annularia*; M. Beeli
Fasc. 3. 1954. Boletineae, P. Heinemann.
Fasc. 4. 1956. *Lactarius*, H. Heim.
Fasc. 5. 1956. *Agaricus* I. and Fasc. 6. 1957. *Agaricus* II and *Pilosace*, P. Heinemann.
Fasc. 6. 1957. *Rhodophyllus*, H. Romagnesi.
Fasc. 7. 1958. *Termitomyces*, R. Heim.
Fasc. 8. 1959. Cantharellineae, P. Heinemann.
Fasc. 9. - 13. incl. non-agaricus.
Fasc. 14. 1965. *Marasmius*, R. Singer.
Fasc. 15. 1966. Hygrophoracerae, *Laccaria* and Boletineae II, P. Heinemann.
Fasc. 16. 1967. Clavaires and *Thelephora*, E.J.H. Corner and P. Heinemann; *Chlorophyllum*, P. Heinemann.
Fasc. 17. 1970. Hydnum s. lato R.A. Maas Geesteranus; *Macrolepiota*, P. Heinemann.
Fasc. 18. 1972. Table des fascicules 1-17., P. Heinemann.

Flore Illust

84

Fasc. 1. 1972. Pleurotoid fungi, D.N. Pegler.

Fasc. 2. 1973. Leucocoprineae p.p., P. Heinemann; *Cystoderma*, P. Heinemann and D. Thoen.

Fasc. 3. 1974. Bolbitiaceae, R. Watling.

Fasc. 4. 1975. *Volvariella*, P. Heinemann.

Fasc. 5. 1977. *Leucocoprinus* and *Aspronocybe*, H. Heinemann and D. Thoen.

Singer, R. and Grinling, K. (1967). Some Agaricales from the Congo. *Persoonia* 4; 355-277. Full descriptions with line-drawings.

West (& East) Africa

Pegler, D.N, (1968 & 1969). Studies en African Agaricales, I. *Kew Bull.* 21 (1968); 499-533. Two *Agaricus* spp., five *Lepiota* s. lato, one *Pluteus*, two Agroscybe (one new), four Coprinaceae, one *Gymnopilus* and sixteen Tricholomataceae (mostly *Marasmius* with three new combinations and five new species by Singer (2) and the author are described. Line-drawings provided. II, Ibid 23 (1969); 219-249. One *Agaricus* sp., six *Termitomyces*, two *Volvariella* (one new) and one each of *Coprinus*, *Hygrophorus*, *Inocybe*, (new), *Lactarius* (new)and *Boletellus*, two *Pleurotus* and four Tricholomataceae (with one new comb.) are described. Line-drawings provided.

Zoberi, M.H. (1972). *Tropical Macrofungi*. London, pp. 158. Descriptions and line-drawings of a very limited selection of agarics. Not recommended.

Zanzibar

Pegler, D.N. (1975). A revision of the Zanzibar Agaricales described by Berkeley. *Kew Bull.* 30; 429-442. Twenty species are considered and supported by four line-drawings.

East Africa

Pegler, D.N. (1977). A preliminary Agaric Flora of East Africa. *Kew Bull. Additional series* (see also *Kew Bull.* 23 (1969); 219-249 and Ibid 23 (1969); 347-412 with R. Rayner--A contribution to the Agaric Flora of Kenya). Full descriptions with line-drawings of both habit sketches and microscopic characters. A most important contribution to the mycology of this part of Africa.

11. MADAGASCAR

Series: *Prodrome Fl. Mycol. Madagascar*. Paris. Full descriptions and line-drawings, Heim, R. (1942). Les Lactario-Russules du domaine oriental Madascar. See under Russulaceae; 1-196. Romagnesi, H. (1941); See Entolomataceae; 1-164. Metrod, G. (1959); see Mycena Tricholomataceae; 1-146.

Also see Heim, R. (1931) Le Genre *Inocybe*. *Encycl. Mycol* 2; 1-249, which covers several Madagascar-taxa in the introduction, including *Tubariopsis* and *Cytarrophyllum*.

12. SOUTH AFRICA

Pearson, A. A. (1950). Cape agarics and Boleti. *Trans. Brit. Mycol. Soc.* 33; 276-316. Six coloured plates and one b/w plate accompany full descriptions. Several new species described; unfortunately no line-drawings are supplied.

Reid, D. A. (1975). Type studies of the Larger Basidiomycetes described from South Africa. *Cont. Bolus Her.* 7; 1-255. A most important re-analysis of type material long since ignored or forgotten. Full descriptions and line-drawings of microscopic characters never before described for these old collections.

South Atlantic Islands

Singer, R. (1955). Agaricales of Tristan da Cunha. *Results Norw. Scient. Exped. Tristan da Cunha.* 38; 15-18. A short account of the species collected on the Norwegian Expedition in 19.

13. NORTH AMERICA

Peck, C. H. (1971-1913). Descriptions of numerous species of fungi, chiefly agarics as well as monographs of many genera. Many coloured illustrations accompany the text. The most important contributions are as follows:

Rep. N.W. St. Bot., see Report 131 (1909) for list of Peck's species and *Mycologia* 54 (1962); 460-465 for species 1909-1915 (R. L. Gilbertson).

33(1880):38 - 49 *Amanita*	35(1882):150 - 164 *Lepiota*
36(1883):41 - 49 *Psalliota*	38(1884):111 - 131 *Lactarius*
38(1884):133 - 138 *Pluteus*	39(1885):58 - 73 Pleurotoid fungi
42(1889):39 - 46 *Clitopilus*	43(1890):40 - 45*Armillaria*
44(1891):38 - 64*Tricholoma*	46(1893):58 - 61*Pluteolus*
46(1893):61 - 69 *Galera*	49(1896):32 - 55 *Collybia*
50(1897):133 - 142 *Flammula*	60(1907):47 - 67 *Hygrophorus*
60(1907):67 - 98 *Russula*	61(1908):141 - 158 *Pholiota*
62(1909):42 - 47 *Lentinus*	
	62(1909):47 - 58 *Entoloma*
63(1910):48 - 67 *Inocybe*	63(1910):67 77 *Hebeloma*
64(1911):77 - 84 *Hypholoma*	64(1911):84 - 86 *Psathyra*
65(1912):59 -89 *Clitocybe*	65(1912):90 - 93 *Laccaria*
65(1912):94 - 105 *Psilocybe.*	

Also note papers on boleti in *Bull. N. Y. St. Mus.* 8. Reprint by Boerhaave Press, Leiden O. O. Box 1051, Netherlands. A very useful drawing together of this important contribution to nineteenth century N. American agaricology (mycology). Descriptions unfortunately lacking microscopic data.

Murrill, W.A. (1910-1934). *North American Flora.* Boletaceae 9 (1910); 133-161. Agaricaceae 9 (1910-1916); 162-426. (Keys and descriptions of genera and species in Cantharellaceae, *Lactarius* and part of Leucosporae). Ibid 10 (1914-1932); 1-348. (remainder of Leucosporae, Rhodo- and Ochrosporae, the latter including *Inocybe* and *Cortinarius* by C.H. Kauffman, and

Pholiota and *Mypodendron* by L.O. Overholts). Many of the descriptions are short and few are supported by microscopic data.

Murrill, W.A. (1912). The Agaricaceae of the Pacific Coast. Series. Parts I - III. *Mycologia* 4; 205-217: 231-262 and 294-308. The first part deals with the Leucosporae, a group completed in part two. Part two also contains descriptions of some brown (pale brown) -spored agarics. Part three contains Melanosporae and the remainder of the Ochrosporae.

Murrill, W.A. (1918). The rosy-spored agarics of North America. *Mem. Brooklyn Bot. Gdn.* 1; 334-336. This paper includes a key to the Rhodosporae.

Kauffman, C.H. (1918). *Agaricaceae of Michigan.* Vol. 1 Text; Vol. 2 Illustrations - b/w photographs. Originally published as the Biol. Series Michigan Geology Biol. Survey Publ. as *Gilled Mushrooms of Michigan and the Great Lake Region.* Reprinted by Johnson, New York (1971). Eight hundred and seventy-five species and varieties are described. Habitat, Time of appearance, etc. are included.

Harper, E.T. (1913-1914). 'Agarics' in the region of the Great Lakes. Series A. *Pholiota. Trans. Wis. Acad. Sci. Arts Letts.* 17 (1913); 470-502. B. *Pholiota* and *Stropharia.* Ibid 17 (1914); 1011-1026. C. *Hypholma.* Ibid 17; 1142-1164. *Lentinus.* Ibid 20 (1921); 365-385.

Beardslee, H.C. and Coker, W.G. (1918-1947), or authors reversed. 'Agarics' of North Carolina. Series. A. *Russula* (Beardslee alone). *J. Elisha Mitchell Scient. Soc.* 33 (1918); 147-199. B. *Lactarius* (Coker alone). Ibid 34 (1918); 1-62. C. Collybias. Ibid 37 (1921); 83-107. D. Laccarias and Clitocybes. Ibid 38 (1922); 98-126. E. Mycenas. Ibid 49 (1924); 49-91. F. *Volvaria.* Ibid 63 (1947); 220-231.

Murrill, W.A. (1922-1923). Dark-spored agarics 'of North America'. Series I. *Drosophila, Hypholma* and *Pilosace. Mycologia* 14 (1922); 61-76. II. *Gomphidius* and *Stropharia.* Ibid 14 (1922); 121-142. III. *Agaricus.* Ibid 14 (1922); 200-211. IV. *Deconica, Atylospora* and *Psathyrella.* Ibid 14 (1922); 258-278. V. *Psilocybe.* Ibid 15 (1923); 1-22.

Smith, A.H. (1933). Unusual agarics from Michigan. *Pap. Mich. Acad. Sci.* 19; 205-216. A whole spectrum of species are described and illustrated. Twelve b/w photographs provided.

Hesler, L.R. (1936-1955). Notes on Southern Appalachian fungi. *J. Tenn. Acad. Sci.* I. 11; 107-122: II. 12; 239-254: III. 16; 161-173: IV. 17; 242-249: V. 18; 290-297: VI. 20; 233-238; VII. 20; 363-372: VIII. 24; 81-93: IX. 26; 4-14: X. 27; 271-277: XI. 29; 205-219: XII. 30; 212-221.
Cover various fungi including agarics and boleti; b/w photographs are supplied with descriptive data.

Smith, A.H. (1938-1941). New and unusual agarics. Series I. *Mycologia* 30 (1938); 20-41. Twenty-two N. American species described oroginally in *Collybia*, two in *Galerina*, two in *Heboloma* (one new), one each in *Psilocybe, Inocybe, Stropharia* and *Tricholoma* and four in *Omphalina* (one new and one transferred to *Galerina*) are considered. Line-drawings and b/w photographs are provided. II. *Mycologia* 33 (1941); 1-16. Ten species, six new mostly *Hypholoma* spp. are considered; in addition three new species of *Inocybe* and one of *Tricholoma*, resembling a white-spored *Inocybe* are considered. B/w photographs are provided.

Smith, A.H. (1941). Study on N. American agarics. *Contr. Univ. Mich. Herb.*

5; 1-73. Fourty-four species mostly new combinations in *Psathyrella* or totally new taxa are described in full. Thirty-two plates of b/w photographs support the text.

Hesler, L.R. and Smith, A.H. (1943). New and interesting agarics from Tennessee and N. Carolina. *Lloydia* 6; 248-266. Ten taxa, eight new are proposed and a key to the varieties of *Collybia maculata* is given. Line-drawings and b/w are offered.

Smith, A.H. (1944). Unusual North American agarics. *Am. Midl. Nat.* 32; 669-697. Twenty-two taxa are discussed over half in the genus *Clitocybe*. Six b/w photographs support the full descriptions; one new combination is also made.

Smith, A.H. and Hesler, L.R. (1946). New and unusual dark-spored agarics from N. America *J. Elisha Mitchell Scient. Soc.* 62; 117-200. Eighteen, all but a few new, are described either in *Psilocybe* or *Psathyrella*. Two species of *Pseudocoprinus* and two of *Coprinus* are described.

Murrill, W.A. (1948-1949). Florida 'agarics'. Series. A. *Lactarius. Lloydia* 11 (1948); 86-98. B. *Amanita.* Ibid 11 (1948); 99-110. C. *Lepiota.* Ibid 12 (1949); 56-61. D. *Tricholoma.* Ibid 12 (1949) 62-69. See also Florida boletes. *Lloydia* 11 (1948); 21-35.

Hesler, L.R. (1957-1959). Notes on South Eastern Agaricales. *J. Tenn. Acad. Sci.* I. 32; 198-207. II. Studies of *Tricholoma* types 33; 186-191. III. Studies of *Collybia* types: 33; 162-166. IV. Studies of *Marasmius* types: 34; 167-171. A very useful series; with b/w photographs in the first part.

Thiers, H.D. (1959). The agaric flora of Texas. I. New species of agarics and boletes. *Mycologia* 49; 707-722. II. New taxa of white- and pink-spored agarics. Ibid 50; 514-523. III. New taxa of brown- and black-spored agarics. Ibid 51; 529-540.

Bigelow, H.E. and Barr, Margaret E. (1960-   ). Contributions to the fungus flora of N.E. North America. *Rhodora* 62--. Covers various fungi including agarics; b/w photographs are supplied with descriptive material.*Rhodora* 62; 186-198. Deals with *Clitocybe, Hygrophorus* and *Lyophyllus*. One new species is described; b/w photographs are offered.

Smith, H.V. and Smith, A.H. (1973). *How to know The Non-Gilled Fungi*, pp. 402, Iowa. Includes the Boletales, Cantharellales, Podoxales, Hymenogastrales and Asterosporales. Some line-drawings with half-tones. Basically a descriptive key, spirally bound for easy use in the field and in the laboratory. For monographic works by these authors see under appropriate generic heading and under popular publications.

Bigelow, H. (1976-   ). Studies on New England agarics I. *Rhodora* 78; 120-134. Includes descriptions of *Amanita, Callistosporium, Collybia* and *Mycena* spp. II. Ibid 80; 404-416. Includes *Stropharia* and *Psilocybe* spp.

Alaska, N. Canada and Northern Rockies

General: *Fungi Canadensis.* Nat. Myc. Herb. Biosystematics Res. Inst. Ottawa, Canada. Plates 2. *Agaricus semotus*; 3. *Inocybe dulcamara*; 31. *Tricholoma fulvum*; 44. *Tylopilus rubrobrunneus*; 113. *Hohenbuehelia longipes*; 133. *Psathyrella typhae*; 142. *Phaeomarasmius rhombosporus*; 165. *Mycena cariciophila*; 166. *Marasmiellus paludosus*.

88

Bigelow, H.E. (1959). Notes on Fungi from Northern Canada. IV. Tricho-lomataceae. *Can. J. Bot.* 37; 769-779. Several combinations are made in *Clitocybe* of fungi formerly placed in *Omphalia*.

Kobayasi, Y. et al (1967). Mycological Studies of the Alaskan Arctic. *Ann. Rep. Inst. Fermentation, Osaka.* 3; 1-138. (agarics and allied fungi: pg. 71-97). Fifty-two taxa are considered; abbreviated descriptions are given supported in some cases by line-drawings. Many collections only identified to genus.

Miller, O.K. (1968). Interesting fungi of the St. Elias Mts., Yukon Territory and adjacent Alaska. *Mycologia* 60; 1190-1203. B/w photographs and line-drawings support descriptions of nine species of agaric newly recorded from Alaska.

Miller, O.K. (1968). Notes on Western Fungi I. *Mycologia* 59; 504-512. Five species, with descriptions and b/w photographs.

Watling, R. and Miller, O.K. (1971). Notes on eight species of *Coprinus* of the Yukon Territory and Adjacent Alaska. *Can. J. Bot.* 49; 1687-1690. Species are described from the St. Elias Mts. and near Kluane Lake in the Yukon Territory the Skolai pass in the Alaskan Range and the vicinity of Juneau, Alaska.

Miller, O.K., Laursen, G.A. and Murray, B.M. (1973). Arctic and alpine agarics from Alaska and Canada. *Can. J. Bot.* 51; 43-49. Nine species are described and supported by line-drawings.

Wells, V.L. and Kempton, P.E. (1975). New and interesting fungi from Alaska. *Nova Hedw., Beih.* 51; 347-358. Four agarics and two boletes, four new, are described; no illustrations are provided.

Arctic

Singer, R. (1954). The cryptogamic flora of the Arctic. VI Fungi. *Bot. Rev.* 20; 451-462. A short account covering the records then known. A useful paper for both agariocologists in Northern N. America and Europe.

Greenland

Lange, M. (1955). Den Botaniske Ekspedition Til Westgrnland 1946. *Medd. Grn.* 147; 1-69. Macromycetes, Part II. Greenland agarics. Descriptions and line-drawings. In English.

Lange, M. (1957). Ibid 148; 1-125. Part III. Greenland agarics continued: 2. Ecological and Plant geographical studies. As above but with b/w photographs.

Kobayasi, Y. (1971). Mycological Studies of the Angmagssalik Region of Greenland. *Bull. Nat. Sci. Mus. Tokyo*, 14; 1-96. (agarics and allied fungi pg. 62-79). Forty-one taxa are considered many only named to genus. Line-drawings support the descriptive data which in many cases are very abbreviated. A useful contribution.

Watling, R. (1977). Larger fungi of Greenland. *Astarte* 10; 61-71. Twenty-three species from forty-seven collections of agarics and puff-balls, four others are assigned questionably to specific names and one collection appears to be a new species of *Lepista*. Several records are the first for Greenland. Descriptive data and line-drawings are offered.

14. TROPICAL N. AMERICA AND CENTRAL AMERICA
See R.W.G. Dennis *Kew Bull. Add. Series* III. see pg. 92.

Murrill, W.A. (1911-1918). The Agaricaceae of Tropical N. America. *Mycologia* 3-5 and 10. To be reprinted as *Bibl. Mycol.* by J. Cramer. See also North American Flora (W.A. Murrill with some genera by C.H. Kauffman and L.O. Overholts) 9 (1907-1916); 1-542: 10 (1-5) (1914-1932); 1-348. Inadequate descriptive data judged by modern standards. *Mycologia* 3 I. *Mycologia* 3 (1911); 23-36. II. Ibid 3 (1911); 79-91. III. Ibid 3 (1911); 189-199. IV. Ibid 3 (1911); 271-282. V. Ibid 4 (1912); 72-83. VI. Ibid 5 (1913); 18-36. VII. Ibid 10 (1918); 15-33. VIII. Ibid 10 (1918); 62-85. I-III inclusive covers Leucosporae, IV covers Rhodosporae. V. and VI. Ochrosporae and VII and VIII. Melanosporae.

Dennis, R.W.G. (1951). Species of *Marasmius* described by Berkeley from Tropical America. *Kew. Bull.* (1961); 153-163. Illustrated descriptions are given: also some Tropical American Agaricaceae referred by Berkeley and Montagne to *Marasmius, Collybia* and *Heliomyces* Ibid (1951) 14; 387-410.

Singer, R. (1945). Farlow's Agaricales from Chocorua. *Farlowia* 2; 39-52. An account of the agarics collected by Farlow in this small area of Mexico.

Singer, R. (1957). Fungi Mexicani, series prima. *Sydowia* 11 (1957); 354-374: Series secunda. *Sydowia* 12 (1958); 221-243.

Guzman, G. (1975). New and interesting species of Agaricales of Mexico. *Nova Hedw., Beih.* 51: 99-118. Ten species, three new, in the Amanitaceae and Pleurotaceae are fully described. Seven plates of line-drawings support text. In addition one new combination in Strophariaceae is made. Many contributions have been made by this author, all in Spanish, and can be sought in *Bol. Soc. Bot. Mex', Bol. Soc. Mex. Mic., An Inst. Biol. Univ. Nal. Auton, Mexico, Ser. Bot.* and *An. Elc. nac. Cienc. Biol. Mex.* Much of this work has been incorporated and condensed in *Identification de los Hongos (Comsestibles, venenosus y alucinantes)* Limusia, 1977. with 425 pp; 582 illustrations and 170 pages of descriptive key.

15. CARIBBEAN

See Dennis, R.W.G. (1970). Fungus Flora of Venezuela and Adjacent countries. *Kew Bull. Add. Series* III This incorporates Dennis' work on West Indian Fungi published earlier in *Kew Bull.* etc. see pg. 92.

Baker, R.E.D. and Dade, W.T. (1951). Fungi of Trinidad and Tobago. *Mycological Paper* 33, Commonwealth Myc. Inst. A list of all the fungi recorded for the area is given including some short descriptions of agarics. Line-drawings and four coloured plates of agarics support the text.

Dennis, R.W.G. (1952). *Lepiota* and allied genera in Trinidad, British West Indies. *Kew Bull.* 9; 459-500. One new *Amanita* and several descriptions of non-lepiotoid fungi are given in addition to a key and descriptions to thirty-four species of *Lepiota* s. lato. Twelve new species are included in the descriptive text; two new combinations and one state nov. are provided along with line-drawings.

Dennis, R.W.G. (1953). Some West Indian collections referred to *Hygrophorus. Kew Bull.* (1953); 253-268. see pg. 92.

Dennis, R.W.G. (1953). Agaricales de l'ile de la Trinite. *Bull. Soc. Mycol. Fr.* 69; 143-198. Descriptions of several species particularly of coloured spored agarics. Three coloured plates offered.

Pegler, D.N. and Fiard, J.P. (1978). *Hygrocybe* Sect. *Firmae* in Tropical America. *Kew Bull.* 32; 297-312. An account of all the members of this section in the New World are given; coloured and b/w illustrations support the text. Several new taxa are proposed.

Pegler, D.N. and Fiard, J.P. (1979). Taxonomy and ecology of *Lactarius* (Agaricales) in the Lesser Antilles. *Kew Bull.* 33; 601-628. Part of a study of the agarics of the Leseser Antilles towards a proposed agaric flora.

16-18. SOUTH AMERICA

GENERAL

Singer, R. (1969). Mycoflora Australis. *Nova. Hedw., Beih.* 29; 1-450. Full descriptions of all known agaricoid fungi recorded for South America, and some related taxa are given. The most important publication on S. American larger fungi but unfortunately lacks illustrations and line-drawings. Incorporates the author's many type-studies. See-Un nuevo hongo comestible de Sudamerica. *Bol. Soc. Arg. Bot.* 10 (1963); 207-208: Two genera new for S. America. *Vellozia* 1 (1961); 14-18. and New agarics from S. America. *Nova. Hedw.* 20 (1970); 785-792. Includes eighteen pages on Gastromycetales and fourteen on larger more conspicuous members of the Aphyllophorales.

Singer, R. in *Flora Neotropica.* 3 (1970)Omphalinae: 4 (1970) *Phaeocollybia*: 5 (1970) Strobilomycetaceae: 10 (1976)Marasmieae. See under individual general headings.

Singer, R. Monographs of South American Basidiomycetes: Series. For full data see under appropriate generic heading.

1. The genus *Pluteus* in South America. *Lloydia* 21 (1958); 195-299.
2. The genus *Marasmius* in S. America. *Sydowia* 18 (1965); 106-358.
3. Reduced marasmoid genera in S. America. *Sydowia* 14 (1960); 258-280.
4. *Inocybe* in the Amazonas region with supplement to Part 1 *Pleutus* 15 (1961); 112-132.
5. Gasteromycetes with agaricoid affinities (Secotiaceae, Hymenogastrineae and related forms). *Bol. Soc. Arg. Bot.* 10 (1962); 52-67.
6. Mesophelliaceae and Cribbeaceae of Argentina and Brazil. *Darwiniana* 12 (1963); 598-611. - with J.E. Wright and E. Horak.
7. The families Paxillaceae, Gomphidiaceae, Boletaceae and Strobilomycetaceae. *Nova Hedw.* 7 (1964); 93-132.
8. Oudemansiellinae, Macrocystidinae, Pseudohiatulinae in S. America. *Darwiniana* 13 (1964); 145-190.
9. *Tricholoma* in Brazil and Argentina. *Darwiniana* 14 (1966); 19-35.
10. *Xeromphalina. Bol. Soc. Arg. Bot.* 10 (1965); 302-310.

Horak, E. Fungi Austroamericani: Series, for full details I, II, & XI see under appropriate generic heading.

I. *Tricholoma* (Fr.) Quelet. in *Sydowia* 17 (1964); 153-163.

II. *Pluteus* Fr. *Nova Hedw.* 8 (1964); 140-199.

III. *Rhodogaster. Sydowia* 17 (1964); 190-192.

IV. Revision de los hongos surperiores de Tierra del Fuego o Patagonia en el Herbario de C. Spegazzini en la Plata. *Darwiniana* 14 (1967); 355-385. A very useful reanalysis of Spegazzini's material.

V. Beitrag zur kenntnis der Gattungen *Husterangium, Hymenogaster, Hydnagium* and *Melanogaster* in Sudamerika (Argentinien and Uruguay). *Sydowia* 17 (1963); 197-205.

VI. Beitrag zur Kenntnis der Gattungen *Martellia, Elasmomyces* and *Cystangium* in Sudamerika. *Sydowia* 17 (1963); 206-213.

VII. *Hypogaea* gen. nov. aus dem *Nothofagus*-wald der patagonische Anden. *Sydowia* 17 (1963); 297-301. This genus is now synonimised with *Setchelliogaster*.

IX. *Gauteria* Vitt., *Martellia* Matt., and *Octavianina* Kuntze in Sudamerika (Chile). *Sydowia* 17 (1963); 308-313.

X. *Crepidotus* Kummer. *Nova Hedw.* 8 (1964); 333-346.

XI. Studien zur Gattung *Thaxterogaster. Nova. Hedw.* 10 (1965); 211-241. with M. Moser. Twenty-eight species are described, the majority new, three subgenera are recognized, two new. Line-drawings support the full descriptive data.

All articles are supported by line-drawings and full descriptions.

Argentina

Singer, R. (1952 and 1953). The agarics of the Argentina sector of Tierra del Fuego and limitrophous regions of the Magallanes areas. I. *Sydowia* (1952) 6; 165-226. The agarics of the Argentina sector. . . II. *Sydowia* 7 (1953); 206-265.

Singer, R. (1953). Quelques agarics nouveaux de l'Argentina. *Rev. Mycol.* 18; 2-23. A preparatory publication for the Prodromo. In French; see below.

Singer, R. and Digilio, A.P.L. (1953). Prodromo de la Flora Agaricina Argentina. *Lilloq* 25; 5-462. An immense work incorporating studies on both fresh and herbarium, often type material. A basic work for anyone contemplating working with agarics from S. America. In Spanish.

Singer, R. (1954). Agaricales von Nahuel Huapi. *Sydowia* 81; 100-157. Descriptive data on the agarics of this area of S. America; all families of agarics are considered. Forty-two species are described, the majority new. List of taxa recorded with *Nothofagus* is given. In German.

Singer, R. (1959). Dos generos nuevos para Argentina. *Bol. Soc. Arg. Bot.* 8; 9-13. Two unusual agarics in the Agaricaceae (*Cystogaricus* and *Volvolepiota*) are described; new name and species proposed.

Brazil

Singer, R. (1955). New species of Agaricales from Pernambuco. *Ann. Soc. Biol. Pernamb.* 13; 225-229. Descriptive data on fungi from this small area of Brazil.

Singer, R. (1965). Interesting and new Agaricales from Brazil. *IMUR Acta* 2; 15-59. An important publication indicating the richness of this vast mycologically poorly explored country.

British Honduras
Smith, A.H. (1939). Notes on Agarics from British Honduras. *Contr. Univ. Mich. Herb.* 1; 21-32. Sixteen species mostly in *Heliomyces*, and many now placed in *Marasmius*, are briefly discussed; one new species of *Mycena* is also described.

Chile
Singer, R. (1959). Basidiomycetes from Masatierra (Juan Fernandez Islands, Chile). *Ark. Bot.* ser. 2, 4; 371-400. Short descriptive data offered.

Singer, R. (1960). Dos especies interesantes Agaricales en Punta Lara. *Bol. Soc. Arg. Bot.* 8; 216-218. One species each of *Psilocybe* and *Pleurotus* are described; line-drawings are provided. In Spanish.

Singer, R. (1960). Three new species of Secotiaceae from Patagonia. *Persoonia* 1; 358-361. Two species of *Thaxterogaster* and one of *Weraroa* are described; a new subgenus is proposed.

Columbia
Singer, R. (1963). Oak mycorrhiza fungi in Colombia. *Mycopath. Myc. Appl.* 20; 239-252. A most important paper giving the first opportunity through an experienced agariocologist's eyes to see the potential richness of the agaric flora of this county.

Ecuador
Singer, R. (1975-1977). Interesting and new species of Basidiomycetes from Ecuador. I. *Nova. Hedw., Beih.* 51 (1975); 239-246. II. *Nova Hedw.* 29 (1977). Reprinted together as single volume. A pair of articles from a previously unexplored country. We welcome further contributions in this series. Descriptive data offered for the first time.

Venezuela and adjacent countries
Dennis, R.W.G. (1970). The fungus flora of Venezuela and adjacent countries. *Kew. Bull. Add. Series* III. Full descriptions with keys and coloured and b/w plates. Supersedes articles appearing in *Trans. Brit. Mycol. Soc.* 34 (1951) 411-482 (dealing with Leucosporae): *Bull. Soc. Mycol. Fr.* 69 (1953); 143-198 (dealing with many Ochrosporae) and in *Kew Bull.* 1951 and 1961-1962 (dealing with Tricholomataceae etc.).

Dennis, R.W.G. (1961-1962). Fungi Venezuelani IV. *Kew Bull.* 15; 67-156. Illustrated descriptions of agarics covering a wide range of genera based on fresh collections and herbarium material. Several new species, some by R. Singer are proposed.

Antarctic
Singer, R. (1956). A fungus collected in the Antarctic. *Sydowia, Beih.* 1; 16-23.

Singer, R. and Corte, A. (1962). Estudio sobre los Basidiomycetes antarcticos. *Contr. Inst. Ant. Arg.* 71; 1-43. The most comprehensive study so far of this continent. In Spanish.

IV. ADDITIONAL REFERENCES

A. GENERAL WORKS
Ainsworth, G.C. (1971). Ainsworth and Bisby's *Dictionary of the fungi.* 631 pp.

Sixth ed., C.M.I. Kew, England. Lists alphabetically all genera often with
useful comments and includes a list of terms; text is supported by line-
drawings. In earlier editions a tabular classification is supplied. Terminology is expanded in *Glossary of Mycology*. Snell, W.H. and Dick, E.A.
(1957). Harv. Univ. Press.

Bessey, E.A. (1952). *Morphology and Taxonomy of the Fungi*. Baltimore, 791
pp.

Hawsworth, D. (1974) The Mycologist's Handbook. Commonwealth Mycological Inst. Kew pp. 230. A very useful publication covering all aspects of
collecting, labelling, processing material and naming, with appendices
on herbarium, collectors names, etc. The nomenclaturial parts suprsede
G. Bisby 'Nomenclature and Taxonomy of the Fungi', published by the
same institute. From the same institute - *The Plant Pathologist's Pocket
book* has great use.

Ainsworth, G.C. and Sussman, A. (1965-1973). *The Fungi*. Series, Academic
Press London. Reference should be made to Vol. 4B - with F.K. Sparrow:
303-504.

B. NOMENCLATURIAL PUBLICATIONS

A) Clements, F.E. and Shear, C.L. (1931). *The Genera of Fungi*, New York 496
pp. Includes agaric genera; superseded by Donk, see below.

Donk, M.A. (1962). The generic names proposed for Agaricaceae *Nova Hedw.*,
Beih. 5; 1-320. All agaric genera, valid and invalid are treated in alphabetical order. Includes information formally published in Nomenclature
Notes on generic names of agarics. *Bull. Bot. Gdn. Buitenz.* Ser. III, 18
(1948); 271-402.

Cooke, W. Bridge (1953). The Genra of Homobasidiomycetes (exclusive of the
Gasteromycetes), spec. publ., Div. Mycology and Dis. Survey, U.S. Dept.
Agr. (mimeographed) 1-100.

Singer, R. and Smith, A.H. (1946). Proposals concerning the nomenclature of
the gill-fungi including a list of proposed lectotypes and genera conser-
vanda. *Mycologia* 38; 240-299. See also by the same authors "Emendations
to a proposal concerning the nomenclature of gill-fungi." Ibid 40 (1948);
627-629.

Horak, E. (1968). Synopsis generum Agaricalium (Die Gattungstypen der
Agaricales). *Beit. zur Kryptogamen - flora der Schweiz* 13; 1-740. Full
descriptions of the genra and type species are given for valid and many
non-valid names.

B) Index of Fungi. Produced by CMI, Ferry Lane, Kew, Richmond, Surrey,
England. Lists names of new genera, species and varieties of fungi (and
from 1971 lichens), new combinations and new names. Petrak's lists
formerly published in Just's *Botanischer Jahresbericht*, 1920-1940 (1935-
1st) have been now reprinted by the Commonwealth Mycological Inst. so
bridging the gap between Saccardo and the first production of the *Index*
in 1941.

C. MYCOLOGICAL LISTINGS
a) Agaric lists
Dennis, R.W.G., Orton, P.D. and Hora, F.B. (1960). New Check List of British

Agarics and Boleti. *Trans. Brit. Mycol. Soc.* 43, Suppl. 1-225. An alphabetical list of all agarics recorded for the British Isles with nomenclature and in the appendix taxonomic notes. Supersedes Pearson, A.A. and Dennis, R.W.G. (1948) Revised list of British Agarics and Boleti. *Trans. Brit. Mycol. Soc.* 31; 145-190. A telescoped source of information on several hundred species of agarics and useful outside the area given in the title.

Miller, O.K. and Farr, D.F. (1975). An index of the common fungi of N. America, synonymy and common names. *Bibl. Mycol.* 44; 1-203. A very handy source of information. See also Konrad, P. and Maublanc, a. (1924-1937). *Revision des Hymenomycetes de France et des pays limitrophes.* Paris, 558 pp. A very useful listing with synonyms, reference to plates etc., although of wider coverage than agarics.

Saccardo, P.A. and others (1882-1928). *Sylloge Fungorum omnium hurscusque cognitorum*, with Vols. 10 and 17 covering bibliography.

Vol. 5. 1887, Agaricineae, Patavii.
   9. 1891, Supplementium Universales with index in Vol. 10 (1892), Patavii.
   11. 1895, Suppl. 3. Patavii.
   12. 1897, Index Universalis (Sydow), Berolini.
   13. 1898, Universalis Matricum conncinnavit (Sydow), Berolini.
   14. 1899, Suppl. 4 (Saccardo and Sydow), Patavii.
   15. 1901, Synonymia (Mussat), Parisiis.
   16. 1902, Suppl. 5, (Saccardo and Sydow), Patavii.
      with cumulative index of genera.
   17. 1905, Suppl. 6 (P.A. Saccardo and D. Saccardo), Patavii with index in vol. 18 (1906).
   19. 1910, Icones Fungorum A-L (Saccardo and Trotter), Patavii.
   20. 1911, Icones Fungorum N-Z (Saccardo and Trotter), Patavii.
   21. 1912, Suppl. 8 (Saccardo and Trotter), Patavii.
   23. 1925, Suppl. 10, (Trotter), Abellini.
   24. 1926, Suppl. 10, (Trotter), Abellini.
   26. 1972, Suppl. (Trotter cord. by E. Cash). Johnson Reprint, New York.
      Sylloge Fungorum has been reprinted by Johnson, New York.

b) Substrate lists:
Oudemans, C.A.J.A. (1919-1924). *Enumerate systematica Fungorum* I-V, *Hagae comitum.*
   See also Saccardo's Universalis Matricum conncinnavit in Vol. 13 Sylloge Fungorum above.

c) Literature lists:
   See Bibliography of Systematic Mycology, Commonwealth Mycological Institute, Kew, Richmond, Surrey, England. See introduction.
Sydow et al. Thesarus Litterature Mycologique et Lichenologique.
Sydow, P. and Lindau, G. (1908-1917). Thesaurus Litteraturae. A-L. (1908), Lipsis.
   Thesaurus Litt. M-Z. (1909), Lipsis: Supplementa 1907-1910 (1913), Lipsis: Argumenta I-IV. (1915) Floristic works, Matrices etc., Lipsis: Argu-

menta V-VIII (1917) Floristic works, Systematics etc., Lipsis.

Cifferi, R. (1957-1960). Thesaurus 1911-1930. A-D. (1957), Papia: Thesaurus E-K (1958), Papia: Thesaurus L-Q (1959), Papia and Thesaurus R-Z (1960), Papia.

See also many articles in *Docums. Mycol.* which cover bibliographic work on the taxonomy and the ecology of selected genera. see pg. 101.

Exsiccatta publications

Lundell, S. and Nannfeldt, J.A. (1934-   ). Fungi Exsiccati Suecici Praesertium Upsaliensis. Uppsala. Descriptive information accompanying herbarium material and available independently as a published account.

Stevenson, J. (1971). An account of Fungi Exsiccati containing material from the Americas. *Nova Hedw. Beih.* 36; 1-563.

Hawksworth, D. (1974). see pg. 93.

D. ADDITIONAL METHODS OF IDENTIFICATION

a) Punch-cards:

Locquin, M. (1968-   ). Une nouvelle methode de taxonomic automatise: *Mycotaxia.* Prof. Locquin has produced several keys for identification, including those for colour recognition, identification of boletes, *Coprinus* spp. etc. Contact direct at Le Village, St. Clement, 89100 Sene, France.

b) Computer identification

Malloch, R. E. and Singer, R. (1971). Bayesian analysis of generic relationships in Agaricales. *Nova. Hedw.* 21; 753-787.

Malloch, R.E. and Singer, R. (1977). Taxonomic position of *Hydropus floccipes* and allied species - a quanitative approach. *Mycologia* 69; 1162-1172.

E. SEMI-POPULAR BOOKS AND ILLUSTRATIVE WORKS

Atlas: as a supplement to *Bull. (Trimest.) Soc. Mycol. France.* A coloured illustration of a taxon or group of taxa with full supporting text, very useful serial publication. See pg. 97.

Bohus, G. and Babos, M. (1978). Coloured illustrations . . . see pg. 97.

Bon, M. (1979). Coloured illustrations . . . see pg. 97.

Cetto, B. (1971). *Funghi dal vero.* Trento, 2nd edition, 619 pp. Lavishly and beautifully illustrated book with coloured plates and macroscopic descriptions. A very useful guide; in Italian.

Dermek, A. and Pilat, A. (1974). *Poznavajme huby veda.* Bratislava, 256 pp. Some fine photographs and reproductions of water-colour paintings.

Dermek, A. (1979). Coloured illustrations . . . see pg. 97.

Groves, J. Walton, (1962). *Edible and Poisonous mushrooms of Canada.* 298 pp., Publ. 112. Research Branch, Can. Dept. Agric. Ottawa.

Haas, H. and Schrempp, H. (1969). *Pilze in Wald and Flur.* Stuttgart 112, 71 pp. Fungi in colour with most descriptions based on macroscopic characters. In German.

Haas, H. and Schrempp, H. (1971). *Zeldzame Paddestochen.* Stuttgart 112, 69 pp. Fungi in colour as above. In German.

96

Hesler, L.R. (1960). *Mushrooms of the Great Smokies*. Knoxville, 287 pp. One
hundred and seventy-eight species of agarics, exclusive of other groups
are dealt with in b/w photographs and text. Keys are offered.

Heim, R. (1957). *Les Champignons d'Europe*. Paris, Two volumes of limited
value for identification but full of interesting biological facts about fungi.
In French. Vol. 1 521 pp.: Vol. 2 236 pp.

Hennig, B. (Michael/Hennig) *Handbuch fur Pilzfreude*. Jena. 1958-1970.
  I. Die Wichtigsten und haufigsten Pilze (1958); 260 pp.
  II. Nich Blatterpilze (1960); 328 pp.
  III. Blatterpilze Hellblatter (1964); 286 pp.
  IV. Blatterpilze Dunkelblatter (1968); 326 pp.
  V. Michlinge/Taublinge (1970); 391 pp.
  VI. Die Gattung der Grosspilze Europeas Bestimmingschlusses und
  Gesamtregister der Bande 1-5 (H. Kreisel), (1975); 290 pp.
See also "Taschenbuck fur Pilzefreunde," Jena (1968), by same author.
Kreisel's work contains very useful keys and systematic arrangements.
The books are illustrated with coloured plates.

Kibby, G. (1979). *Mushrooms and Toadstools, a field guide*, Oxford, 256 pp.
Coloured illustrations of more than four hundred British fungi, many
never previously illustrated, are given with key, commentary and descriptions.

Marchand, A. (1971-1977). *Champignons du nord et des midi*.
Perignon. At present in five volumes. Beautifully coloured illustrations
accompanied by field descriptions; microscopic characters are added as
additional appendix. In French.

Miller, O.K. (1973). *Mushrooms of North America*. New York, 224 pp. Covers
680 species found in continental U.S.A. and Canada. Short descriptions
supported by coloured photographs, line-drawings and key.

Nilsson, S. and Persoon, O. (1977). *Svampar naturen*. Illustrations by B. Moss-
berg. 2 volumes, Stockholm, 127 and 131 pp. respectively. Excellent
illustrations and ecological data supporting the mainly macroscopic data.
In Swedish. Republished as *Fungi of Northern Europe* Vol. 1. Larger
Fungi and non gill-Fungi and Vol. 2 Gill-Fungi, translated by David
Rush and edited and adapted by David Pegler and Brian Spooner, 1978.
Excellent for introductory information.

Persson, O. (1971). *Matsvampar i farg*. Stockholm, 142 pp. Although restric-
tive, excellent for the ecological data particularly photographs of collecting areas similar to those of Fries.

Pilat, A. and Usak, O. (1958). *Mushrooms*. 220 pp., London. Full macroscopic
descriptions support the faithfully reproduced watercolour paintings.

Pilat, A. and Usak, O. (1961). *Mushrooms and other Fungi*. London, 320 pp. A
companion to the 1958 book and printed in the same way. Also obtainable
as a pocket edition.

Printz, P. (1976). *Forums bog on Swampe*. Copenhagen, 143 pp. Ninety-seven
coloured photographs, eighty of agarics, with short macroscopic descriptions. In Danish.

Reid, D.A. (1966). Coloured illustrations of rare and interesting fungi. . . see
pg. 97.

Rinaldi, A. and Tyndale, V. (1974). *Mushrooms and other fungi*. London, 333 pp. Well illustrated with reproductions of water-colours; short field descriptions are supplied but the book is unusual in its class in that it shows the variation found in Europe in a single taxa. First published in Italian (1972).

Romagnesi, H. (1956-67). *Nouvelle Atlas des Champignons*. 4 Vols. Bordas, Paris. Also issued in 3 small volumes as *Petite Atlas des Champignons* (1967). A bulky work not covering very many taxa. Good illustrations with useful supporting text. In French.

Smith, A.H. (1949). *Mushrooms in their Natural Habitats*. Portland 626 pp. accompanied by a set of coloured-slides and viewer. The text has been reprinted in 1973 (Haynes, New York.).

Smith, A.H. (1963). *The Mushroom Hunters' Field Guide*. Revised and enlarged, 264 pp., Ann Arbor. First produced in 1938. 187 coloured plates and more than two hundred b/w photographs of common North American Fungi are given. The illustrations are supported by keys and telescoped descriptions.

Smith, A.H. (1975). *A Field Guide to Western Mushrooms*. Ann Arbor, 278 pp. A companion to the above. Two hundred and one species are dealt with and supported by excellent colour photographs.

Stuntz, D. (1971). *The Savory Wild Mushroom*. London, 242 pp.. A revised and enlarged version of Margaret McKenny's book originally produced in 1962. The book is arranged in much the same way as Hesler's *Mushrooms of the Great Smokies* and Smith's Field Guides. Sixty-five coloured plates support the text.

Suber, Nils (1968). I Svamp skogen, Uddevalla, 223 pp. Short descriptions with line-drawings, b/w photographs and thirty-two coloured plates covering eighty species. In Swedish.

Wakefield, E.M. and Dennis, R.W.G. (1951). *Common British Fungi*, London, 290 pp. Well-illustrated descriptions covering the majority of large fungi encountered in England. One hundred and eleven coloured plates depict these fungi and comments are included on additional species. Now a collector's piece.

Reid, D.A. et al (1966-   ): Reid 1966-1969 and 1972; E. Schild 1971; M. Moser, 1978; G. Bohus and M. Babos, 1977; S.P. Wasser, 1979; A Dermek, 1979; M. Bon, 1979.

  I. *Boletus, Laccaria, Lepiota, Amanita* and *Stropharia*, 1966.

 II. *Lepiota, Tricholoma, Porpoloma, Nolanea, Melanoleuca, Pluteus, Hypholoma, Naucoria* and *Lactarius*, 1967.

III. *Boletus, Hygrophorus, Lepiota, Nolanea, Clitocybe, Lyophyllum, Mycena, Rhodocybe, Pluteus, Conocybe, Pholiota, Psathyrella and* Heboloma, 1968.

IV. *Clavulinopsis, Russula* and *Lactarius*, 1969.

 V. *Ramaria, Clavulinopsis*, and *Clavaria*, 1971 (Schild).

VI. *Hygrocybe, Hygrophoropsis, Agaricus, Lepiota, Leptonia, Pluteus, Coprinus, Tubaria, Inocybe, Cortinarius, Russula, Stropharia, Gyrdon, Russula* and *Phaeocollybia*, 1972.

VII. *Hygrophorus, Hygrotrauma, Tricholoma, Porpoloma, Mycena, Pholiota, Stropharia, Conocybe, Pachylepyrium, Bolbitius, Melanotus, Agrocybe, Simocybe, Naucoria, Xeromphalina, Lactarius* and *Inocybe*, 1978 *(Moser)*.

VIII. *Agaricus, Armillaria, Coprinus, Inocybe* and *Leucopaxillus*,
   1977 (Bohus and Babos).
 IX. Boletaceae: *Boletinus, Leccinum, Buchwaldoboletus, Xerocomus, and Boletus* sensu stricto, 1979 (Dermek).
  X. *Agaricus, Cystoderma, Leucocoprinus, Leucoagaricus* and *Galeropsis*,
   1979 (Wasser).
 XI. *Hygrocybe, Lactarius, Russula, Calocybe, Heleloma, Inocybe, Leucoagaricus, Lepiota*, and *Amanitopsis*, 1979 (Bon).
   A continuing series of illustrations of larger fungi in coloured and supported by full descriptions and line-drawings. Many of the fungi depicted are new species or taxa which are rarely met. At the moment confined to European fungi but of great use to any worker in the tempeture zone.
   This important series is published by J. Cramer, Vaduz. See *Nova Hedw.*

F. CLASSIC TEXTS

   The nomenclaturial starting point for agaric taxonomy is taken as *Systema Mycologicum* (1821) written by the 'Father of Myclolgy', Elias Magnus Fries.

1. Pre-starting point books in alphabetical order of authors.

*Albertini, J.B. de and Schweinitz, L.D. de (1805). *Conspectus Fungorum in Lusatiae superioris agro Niskiensi crescentium*, Lipsiae, 376 pp..

Batsch, A.J.G.C. (1783-1789). *Elenchus Fungorum*, Halae; 183 pp. and 271 pp. + 163 pp. (1889).

Bolton, J. (1788-179). A *History of Funguses growing about Holifax*. Halifax and Huddersfield, 3 Volumes and Supplement. (For account of this author see Watling, R. and Seaward, M. in *J. Bibl. Nat. History*, 1980).

Bulliard P. (1780-1793). *Herbier de la France*, Paris, including *Histoire des Champignons de la France* (1791-1812), Paris.

Persoon, C/ H. (1796). Observationes *Mycologicae, Lipsiae*, 115 pp., reprinted by Johnson, New York.

Persoon, C.H. (1801). *Synopsis Methodica Fungorum*, Gottingae, 706 pp.. Includes details outlined in *Tentamen dispositionis methodicae Fungorum* (1797).

Schaeffer, J.C. (1762-1774). *Fungorum qui in Bavaria et Palatinata circa Ratisbonam nasuntur Icones*, Erlangae, 136 pp., 330 plates. Includes commentary by Persoon.

2. Publications of E.M. Fries

Systema Mycologicum I, Lundae, 520 pp.. Incorporates observations made by the same author in Mycol. Obs. published in 1801.

Elenchus Fungorum sistens commentarium, in *Systema Mycologicum*, printed under the title *Systema Mycologicum ... suppl. voluminis primi*, Gryphiswaldieae, 238 pp.. *(1821).

Epicrisis Systematicus Mycologici (1836-1838), Upsaliae, 608 pp.. *Summa vegetabilium Scandinaviae Section Posterior.* (1849), Holmiae and Lipsiae, 572 pp..

Monographia Hymenomnycetum Sueciae Vol. 1. sistens agarics, Coprinos, Bolbitos, 1857, Vol 2. 1863 Cortinarios etc. and commentary on Icones

Hymen., Upsaliae.

Hymenomycetes Europaei *(1874), Upsaliae, 755 pp.. Reprint available by Asher, Amsterdam (1963).

3. Classic works: post Friesian publications

*Bataille, F. (1902). *Flore monographie . . . d'Europe*. I. *Amanita* and *Lepiota*, 1902. II. Asterospores, 1908. III. Les Boletes, 1908. IV. Cortinaires, 1912. V. Hygrophores 1910. IV. Inocybes, 1910, Besancon and *Bull. Soc. d'Enulation du Doubs*.

Berkeley, M. (1860) etc. *Outlines of British Fungology*, see this author also with C. E. Broome, 1879. *Ann. Mag. Nat. Hist.* 1838-1889, with W.A. Curtis. . . etc. *Notices of British Fungi* are available as single reprint volume by J. Cramer. *Notices on N. America Fungi* (1872-76) has been reprinted by Linnacus Press, Amsterdam, (1971), 1-1005.

Bresadola, G. (1881-1900). *Fungi Tridentini*. Tridentini, 106 pp.. Descriptions and illustrations of many larger fungi particularly agarics. Reprint now available.

Bresadola, G. (1927-1933). *Iconographia Mycologia*, Milan, Twenty-eight volumes. Descriptions supporting coloured illustrations covering a wide range of large fungi. Vol. 27 represents Gilbert's study of *Amanita*; see pg. 7.

Cooke, M.C. (1880-1890). *Illustrations of British Fungi*. London. Eight volumes without descriptive data. 1, 198 coloured plates are provided. A valuable work especially as it indicates the species-interpretations at the turn of the century. A commentary on the illustrations by A.A. Pearson, L. Quelet and R. Maire appears in *Trans. Brit. Mycol. Soc.* 20 (1936); 33-95. Despite the title only agarics are considered.

Britzelmayr, M. (1879-1894). *Die Hymenomycetes von Sudbayeren*. Berlin. An often forgotten, although most important work illustrated with numerous coloured plates. *(Ber. d. naturhist. Var. in Augsburg)*.

Eable, f. S. (1909). The genera of the North American Gill-fungi. *Bull. N.Y. Bot. Gdn.* 5; 373-451. A most important publication from the nomenclaturial point of view. Earle also published keys to various genera based mainly on literature information; these appear in *Torreya* 2-3 (1902-1903).

Fayod, V. (1889). Prodrome d'une histoire naturelle des agaricinees. *Ann. Sci. Nat. Bot.* 7, series 9; 181-411. Probably one of the most single influential publications on modern thought on agaric-classification and although forgotten for many generations sets the scene of the present approach to agaric systematics.

Fuckel, L. (1869-1875). *Symbolae Mycologicae*, Wresbaden. *(Jahrb. Nassauischer Verein Natdke.)* Reprinted by J. Cramer (1966), together with supplements to *Symbolae . . .* I-III.

Gillet, C.G. (1874-1878). *Hymenomycetes de France*, Paris. A very important illustrated account on larger fungi.

Kalchbrenner, C. (1873-1877). *Icones selectae Hymenomycetum Hungariae*, Pestini. A classic, important but rare series of coloured illustrations.

Karsten, P.A. (1882). Rysslands, Finalnds och den Skandinaviska Hattsvampar. *Bidrag till Kannedom af Finland Natur Och Folk.* 32 (1879); 1-571 & 37 (1882); 1-257. A classic work, tabulating the majority of agarics then known from Northern Europe. Many new genera are proposed; in German.

Konrad, P. and Maublanr, A. (1924-1935). *Icones selectae Fungorum*, Paris. A standard work.

Krombholz, J. von (1831-1846). *Naturgetreue Abbildungen und Beschereibungen der Essbaren Scadlichen und Verdachtigen Schwamme*, Prague.

Kuhner, R. (1926). Contribution a l'etude des Hymenomycetes et specialement des Agaricacees. *Botaniste* 17; 1-224. A fundamental examination of the agarics based on the author's own studies. Accurate work illustrated with line-drawings and b/w photographs in a fashion we have come to expect from this author.

Kuhner, R. (1977-   ). Les Grande Lignes de la classification des Agaricales, Asterosporales et Boletales survol histoire et critique. *Bull. Soc. Linn. Lyon* 46 -. A most important analysis of the classification of agaricoid fungi resulting in a rather different approach to that of Rolf Singer; chemical, physiological, cellular data are all included in the analysis. A continuing series with nomenclaturial implications promised in the future.

Lange, J.E. (1935-1939). *Flora Agaricina Danica*. Five volumes, Copenhagen. Each volume with forty coloured plates. First published under the joint auspices of the Society for the Advancement of Mycology and the Danish Botanical Society, Copenhagen. A most important work of importance outside the confines of Denmark. Vol. 1 Leucosporae 90 pp. Vol 2. Leucosporae & Rhodosporae 105 pp. Vol 3. *Cortinarius, Pholiota, Inocybe* and *Hebeloma*, 96 pp. Vol 4. Concludes Ochrosporae; Melanosporae. 119 pp. Vol 5. Hygrophoraceae, Russulaceae, Cantharellaceae and additions. 100 pp.

Massee, G. (1892-1893). *British Fungus Flora. A classified Text-book of Mycology*, London. Three volumes totally or in part dealing with agarics. Now totally superseded, but similar in format to *European Fungus Flora: Agaricaceae* (1902), by the same author which includes the descriptions of agaric-taxa long since forgotten. Vol. 1. *Boletus* s. lato, 258-300: Melanosporae, 300-418. Vol. 2 460 pp. Vol. 3 Leucosporae 1-268.

Patouillard, N. (1883-1889). *Tabulae Analytique Fungorum*, Poligny and Paris. One of the three important, almost revolutionary, publications by this author; 1-700.

Patouillard, N. (1900). *Essai Taxonomique sur les families et les genres des Hymenomycetes*, Lons. la Saunier, 184 pp..

Patouillard, N. (1922). *Les Hymenomycetes d'Europe*, Patis. 166 pp..

Persoon, C.H. (1822-1828). *Mycologia Europae*, Erlangae, 214-282 pp.. A fundamental work by an author who was very discerning in his observations. see pg. 98.

Quelet, L. (1872-1876). *Les Champignons du Jura et Vosges*, Paris: *(Mem. de la Soc. d'emulation de Monteliard)*.

Quelet, L. (1886). *Enchiridion Fungorum in Europa media et praesertium in Gallia Vigentium*. Lutetiae, 352 pp. (See also L'Enchiridion .... *Bull. Soc. Mycol. Fr.* 1; 1-134).

Quelet, L. (1888). *Flore mycologique de la France et des pay limitrophes.*

Rabenhorst, L. (1884).*Kryptogamenflora von Deutshland, Osterreich und der Schweiz*, Vol. 1 Leipzig. 459-864. A series of volumes, only the first of interest to agaricologists. Reprinted by Johnson, New York (1971).

Ricken, A. (1915). *Die Blatterpilze (Agaricaceae) Deutschland und der angrenzenden Lander besonders Oesterreich und der Schweiz,* Leipzig. 480 pp., with 112 coloured plates. Full descriptions in German support the illustrations. Still one of the foremost European publications and although Ricken often adopted a quite personal interpretation of many Friesian taxa his descriptions are so good for this time correct identification can be achieved. The same author also published *Vadecum fur Pilzfreunde* (1920) Leipzig 352 pp., of which a reprint is now available by J. Cramer (1969).

Roumeguere, M. C. (1880). *Flore Mycologique du Dept. de Tarn et Garonne: Agaricinees*, Montauban. 278 pp..

Schroeter, J. (1885-1908). *Die Pilze Schlesiens*, Breslau. Two volumes: Kyrptogramenflora von Schlesien III, 1 - 2. Reprinted as *Bibl. Mycol.* 34a (1974) and 34 (1972).

Schweinitz, L.D. von (1822). Synopsis Fungorum Carolinae superioris secundum observationes L. de Schweinitz. (*Schriften Naturf. Gesell. Lepzig,* Band 1.), Leipzig. Reprinted as *Bibl. Mycol.* 49. (1974); 1-122.

Schweinitz, L.D. von (1834). Synopsis fungorum in American borealis media degentium *(Trans. Am. Phil. Soc.* 4.) Philadelphia 216 pp.. Reprinted in 1962 by J. Cramer.

V. JOURNALS AND PERIODICALS

A. MYCOLOGICAL ORGANIZATIONS

A much more complete list of mycological journals has been compiled by Dr. C. W. Ainsworth and was issued on the occasion of the first International Mycological Congress at Exeter in 1971 as a separate publication. Those journals listed here have articles on larger fungi. Those titles in italics appear in the text; the amount of italics indicates the abbreviations used. For further explanation see pg. 104-110.

Argentina
Matrodiana 1970-. J. Raithelhuber, Int. Alfaro, 179 Acassuso, Buenos Aires, Argentina formerly (editor Fleischfresser, Stuttgart I, Urbanstasse 69, DBR).
Now published from Germany; see under that country. In German.

Austria
Sydowia 1947 -. Continuation of Annales Mycologici (1903-1943). E. Horak, Institut f. Spezielle Botanik, Eidgen. Techn. Hochschule, Universitatstrasse 2. CH 8092, Zurich, Switzerland.

Belgium
Sterbeeckia (Antwerpse mycologische Kring) 1961 - Astridplein 26, B-200, Antwerpen. In Belgian.

Canada
Le Cercle des Mycologiques amateurs de Quebec. 1956 Pavillon des sciences

Purs, Cite Universitaire Ste-Foy, Quebec. In French.

Czechoslovakia

Ceska Mykologie 1954 -. Ceskoslovenska vedecka spolecnost pro mykologii. (Czech Sci. Society for mycology) Continuation of *Mykologia* 1924 - 1953. In Czech. Krakovska 1, Praha 1, (editor: Narodni Mus. Vaclavske nam. 68, Prague).

Mykologicky sbornik (Casopis ceskoslovenskych Houbaru) (Csekoslovenska mykologicka spolecnost - Czech Mycological Society) In Czech. Karmelitska 14, Praha, 1.

Denmark

Friesia (Nordisk Mycologisk Tidsskrift) 1931 -. Foreningen til Swampekundskaben Fremme. Articles often in English: notes in Danish. F. Buchwald, Dalgas Boulevlard 68, Copenhagen, DK - 2000.

Finland

Karstenia 1950 - Suomen Fienseura-Finlands Svampvanner: Finnish Mycological Society. In English, Finnish and German. Editor: Forest Res. Institute, Unioninkatu 40, 00170 Helsinki 17 (associated publication: Sienitietoja-Svampnytt)

France

Bulletin Trimestrial de la Societe mycologique de France. 1885 -. official organ of the French Mycological Society. In French. Editor. 36, rue Geoffroy-St-Hilaire, Paris 53.

Documents Mycologique 1971 -. Mycological papers of an ecological nature. In French. Prof. J.M. Gehu, Lab. de Systematique et l'Ecologique Vegetales, Univerite de Lille II, rue Laguesse 59, Lille.

Bulletin de la Federation Mycologique Dauphine - Savoie. Articles in French. Federation Mycologique Dauphine-Savoie Foey rural de Montmelian, Savoie.

Revue de Mycologique 1936 -. Continuation of Ann. Cryptogamie exotique. Official organ of Lab. de Cryptogamic, Museum National du Histoire naturelle, rue de Buffon, Paris V. Includes Memoire-lors-series. In French.

Germany

Boletus Kulturbund der Deutschen Demokratischen Republik 1977-. Zentral Kommission Natur und Heimat Zentraler Fachausschuss Botanik. Arbeitsmaterial mykologie. D. Dorfelt, Martin-Luther University, Botanic Garten 402 Halle/S Neuwerk 21, In German.

Mykologisches Mitteilungsblatt. 1957- An amatuer club. Frau Mila Herrmann, 402 Halle (Salle), Marthastrasse 27, DBR. In German.

Sudwestdeutsche Pillzrundshau. Replaces earlier Stuttgarter Pilz-Rundschau. 1965 - In German. Mainly amateur society. H. Steinmann, 7305 Altbach/Neckar, Wilhelmstrasse 22. In German.

Westfalsche Pilzbrief. 1958 - . In German.

H. Jahn, 4931 Heilingenkirken/Detmold, Altersportplatz 466, DBR.

Also see Argentina: L Metrodiana now from J. Raithelhuber, Triebweg 199, D 7000 Stuttgart 30 DBR.

*Zeits*chrift fur *Pilzk*unde 1922- In German Deutsche Gesell-
schaft fur Pilzkunde, Wilhelmstrasse 22, D 7305 Altbach/Neckar. Now
replaced by Zeitschrift Mycologie. See pg. 109.

Hungary

Milkologiai Kozlemenyek. In Hungarian. Szabadsag ter. 17 Budapest V (available for exchange only)
Acta Mycologica Hungariae (Magyar Gombaszati Lapok).

Italy

Rassegna Micologiea Ticinese. 1968 -. In Italian.
Societa Micologiea Carlo Benzoni, Chiasso.
Boletino del Gruppo micologico G. Bresadola, 1957 -. In Italian. Gruppo
Micologico G. Bresadola, Museo Tridentino Scienze Naturali, Casella
Postale 168, Trento.
Micologia Italiana. 1972 - In Italian, sometimes English summaries. Gruppo
Giornalistico Edagricole Bologna, Via Filippo Re 8, 40126 Bologna.

India

Indian Journal of Mycological Research , 1955 -. In English. Official organ of
the Indian Mycological Society. Dept. of Botany, University College of
Science, Calcutta.
Kawaka 1975 -. In English
Official organ of Mycological Section, Department of Advanced Botani-
cal Studies, University of Madras, Madras.

Japan

*Trans*actions of the *Myco*logical *Society* of *Japan.* 1956 -. In English and
Japanese. Official organ of the Mycological Society of Japan; National
Science Museum Veno Park, Tokyo.
Nagaoa: Mycological Journal of Nagao. 1952 -. Articles in English and
Japanese. Nagao Institute, Mistiuki - Machi 380 - 21, Setagaya-ku. Tokyo.

Mexico

*Bol*etin de la *So*ciedad *Me*xicana de *Mic*ologia 1968 -. In Spanish, sometimes
with English summaries. Official organ of the Sociedad Mexicana de
Micologia: G. Guzman, Departamento de Botanica, Instituto Politechnico
Nacional, Apartado Postal 42-186, Mexico 17 DF.

Netherlands

Coolia 1954 -. In Dutch, Nederlandese Mycologische Verneeniging: Meg M.
Dijkstal Dukatendreefs, Cuijck a/d Maas. (This society has published
several journals including Medeelingen van de Neerl. 1910-1952, Fungus
1929-1933 and for *Meded. Ned. Mycol. Vereen.* 1933-1958.).
Mycopathologia 1938-1947: (superseded by *Mycopath*ologia et *Mycol*ogia
*appl*icata 1951-1975) now Mycopathologia Mycological Journal, published by W. Junk N.V., The Hague, Netherland.
Persoonia 1959. In English. Official organ of the Mycology Section,
Rijksherbarium Schelpenkade, 6 Leiden.

104

Poland
Acta Mycologica 1965 -. Polskie Towarzytwo Botaniczne: A. Skirgiello, N. Ujazdowskie 4, Warszawa. In Polish, summaries in English.

Sweden
Goteborgs Svampklubb, Arsskrift. 1971-. In Swedish and English. Official organ of the Mushroom Club. K. G. Ridelius, Aschebergsgaten 9, 411 27, Goteborg.

Switzerland
Schweizerische Zeitschrift fur Pilzkunde: Bulletin Suisse de Mycologie 1923 - Official organ of the Swiss mycological Society: Adolf Nyffenegger Muristrasse 6, 3123, Belp. Articles mostly in German; some in French.

United Kingdom
Mycological Papers 1925-. Official organ of the Commonwealth Mycological Institute, Ferry Lane, Kew Richmond, Surrey, England.
Transactions of the British Mycological Society. 1897 - (British Mycological Society) G. Pegg, Wye College, Kent, England (associated *Bulletin*).

United States
McIlvainea 1972- (continuation and up-grading of Mycophile) Official organ of the North American Mycological Association: H.S. Knighton, 4245 Redinger Road, Portsmouth, Ohio 45662. Often incorporates notes in Mycena News (San Francisco Myc. Soc.), etc.
Mycologia 1909- (continuation of Journal of Mycology 1885-1908.) Official organ of the Mycological Society of N. America: H. D. Thiers, Dept. of Botany, San Francisco State University, Holloway Ave., San Francisco. (associated publication; Mycologia Memoirs and Mycological Society of America Bulletin). Journal of Mycology has been reprinted in entirety by Johnson, New York.
Mycotaxon 1974 -. Independently published journal. P.O. Box 264, Ithaca, New York, 14850. Several small societies have sprung up in the United States and some produce their own News-sheets or even small journals e.g. Boston Mycological Club, Oregon Mycological Society, San Francisco Mycological Club.

United Soviet Socialist Republic
Estonia: Mycological Section of the Naturalists' Society. Academy of Sciences of the Estonian S.S.R., 3 Hariduse St. Tartu (A. Raitviir). Intimately connected with the production of Folia Cryptogamica Estonica. Ukraine: Mycological Section of the Ukrainian Botanical Society, 4 Repina St., Kiev (S.P. Wasser). There does not appear to be a mycological society as such in U.S.S.R. The Botanical Society of the USSR has a Mycological Section: 2 prof. Popova St., Leningrad (B.P. Vassilkov). See *Mikologija i Fitopatologija*, from the Academy of Sciences, Leningrad.

B. PERIODICALS AND JOURNALS: ABBREVIATIONS
The journals cited in this work are listed below in full with their

abbreviations as set out in the World List of Scientific Periodicals. However in keeping with the most recent issues of the List, no lower case letters are used in many of the names journal titles. Where the abbreviation used deviates from the World List the correct abbreviation is indicated. Some important serial books are also included in the list below. A few titles do not occur in the World List.

Acta Botanica Islandica, 1972 - *Acta Bot. Isl.*.

Acta Botanica Indica, 1973 - *Acta Bot. Ind.*.

Acta Mus. nat. Prague. see *Sbornik - Praze.*

Acta Mycologica Hungariae see *Magy. gomb.Lap.* below.

Agricultural Gazette of New South Wales, 1890 - *New Agric. Gaz.N.S.W.*

American Journal of Botany, 1914 - *Am. J. Bot.*.

American Midland Naturalist, 1909 - *Am. Midl. Nat.*.

Annals of Botany, London 1950 - *Ann. Bot.* (London), includes *Mem*oirs.

Annales Mycologici, 1903-1944. *Annls.Mycol.* continued as Sydowia. see pg. 108.

Annales historico-naturales Musei nationalis Hungarica, 1903 - *Annls. hist.nat. Mus. natn. Hung.*.

Annales Societe Linneenne de Lyon, 1826-1887. *Annls. Soc. Linn. Lyon.*

Annual Report of the Institute of Fermentation, Osako. *Rep. Inst. Ferm., Osako.*

Astarte, 1967 - unchanged.

Atlas Champignons de l'Europe, Prague. Series of volumes covering larger fungi. See individual generic entries.

Atlas mycologique, a series of books published by P. Lechevalier, Paris, 1964 - abb. *Atlas Myc.* (I. Psalliotes: II. Les Bolets).

Australian Journal of Botany, 1953 - *Aust. J. Bot.*.

Beihefte Botanischen Centralblatt, 1891-1943. *Beih. Bot. Zbl.*

Bericht der Schweizerischen Botanischen Gesellschrift, 1891 - *Ber. Schweiz Bot. Grs.*

Bibliography of Systematic Mycology; see introduction.

Bibliotheca Mycologica. Reprints and original works published by J. Cramer, Vaduz.

Boletin de la Sociedad Argentina de Botanica, 1945 - *Boln. Soc. Arg. Bot.*

Boletin de la Sociedad Mexicana de Micologia, 1968 - *Bol. (Inf.) Soc. Mex. Mic.*.

Botanical Magazine, Tokyo, 1887 - *Bot. Mag., (Tokyo).*

Botaniste, 1888 - unchanged.

Botanisk Tidsskrift, 1866 - *Bot. Tidssk.*

The Botany of Iceland, 1912 - *Botany Icel.*

Botanical Review, 1935 - *Bot. Rev.*

Brooklyn Garden Memoirs, see Memoirs below.

British Fungus Flora: Agarics and Boleti. A serial publication ultimately covering all species recorded for the British Isles. Her Maj. Stat. Office U.K.

Brittonia, 1931 - unchanged.

Bulletin of the British Mycological Society, see pg. 104.

Bulletin de l'Institut agronomique et des stations de recherches de Gembleux, 1932 - *Bull. Inst. agron. Stns. Rech. Gembleux.*

Bulletin du Jardin Botanique de l'Etat Bruxelles, 1902-1941, which became Bulletin du Jardin Botanique Belgique. 1941 - *Bull. Jard, Bot. Nat. Belg.*

Bulletin du Cercle des Mycologiques Amateurs de Quebec, 1956 - *Bull. Cercle Mycol. Amateurs, Quebec.*

Bulletin of the New York State Museum, 1887-1955. *Bull. N.Y.St. Mus.*

Bulletin of the Research Council of Israel. Sect. D. Botany. Bull. Res. Council Israel, 1951-1968; see *Israel J. Bot.*

Bulletin de la Societe Botanique Suisse 1891 - *Bull. Soc. Bot. Suisse,* see under Bericht. . .

Bulletin mensuel de la Societe Linneenne de Lyon, 1932 - *Bull. (mens.) Soc. Linn. Lyon.*

Bulletin de la Societe des naturalistes d'Oyonnax pour l'etude et la diffusion des sciences naturelles dans la region. 1947 - *Bull. Soc. Nat. Oyonnax.*

Bulletin, Societe royale de botanique de Belgique, 1862 - *Bull. Soc. r. Bot. Belg.*

Bulletin Trimestriel de la Societe Mycologique de France, 1903 - *Bull. (trim.) Soc. Mycol. Trim.* is ommitted as the Society's Bulletin ceased in 1903.

Bulletin Suisse de Mycologie see pg. 104.

Cahiers de la Maboke, 1963 - Cah. Maboke

Canadian Journal of Botany, 1951 - *Can. J. Bot.*

Cavanillesia, 1928 - 1938. unchanged.

Ceska Mykologie, 1947 - *Ceska Mykol.*

Collectanea botanica a Bareinonensi Botanica Institute edita, 1946 - *Collnea. bot.,* Barcinone.

Contributions from the University of Michigan Herbarium, 1939-1942. *Contr. Univ. Mich. Herb.*

Coolia, 1954 - unchanged; see pg. 103.

Dansk botanisk Arkiv. 1913 - *Dansk Bot. Ark.*

Darwiniana, 1922 - unchanged.

Documents Mycologiques, 1971 - *Docums. Myc.* see pg. 102.

Encyclopedie Mycologique; a series of books published by P. Lechevalier, Paris 1931 - abb. *Encycl. Mycol.* Volumes I, VII, X, XIV, XX, XX and XXXII are of interest to agaricologists.

Etude Mycologique; a series of books published by P. Lechevalier, Paris 1962 -   . abb Etude Mycol. I. Les Bolets, II. Les Cortinaires, III. Les Lactaries.

Folia Cryptogamia Estonica, 1969. *Fol. Crypt. Est.* See pg. 104.

Farlowia, 1943-1955. Unchanged.

Flora Neotropica, 1968-   . Monographs published by the New York Botanic Gardens, Bronx. Intends ultimately to cover all groups, larger fungi included; abb. *Flora Neot.*

Flore Iconographia des Champignons du Congo, 1935-1972; abb. *Flore Icon. (Champ. Congo).* See below.

Flore Illustree des Champignons d'Afrique Centrale, 1972 - replaces

Flore Icon. A series covering various groups of larger fungi based on the collections of Madame Goosens-Fontana; abb. *Flore Illust. Champ. Afr. Cent.*.

Friesia, 1932- . Unchanged see pg. 102.

Fungus. 1929-1933. Unchanged see pg. 103.

Hedwigia, 1852-1944. Unchanged: replaced by *Nova Hedwigia* see below.

Indian Mushroom Journal, 1975. - *(Ind. J. Mushrooms)*.

Israel Journal of Botany, 1969 - *Israel J. Bot.*. A continuation of Bull. Res. Israel.

Journal of Botany, 1863-1942. *J. Bot., Lond.*

Journal of the Elisha Mitchell Scientific Society, 1883 - *J. Elisha Mitchell Scient. Soc.*

Journal of the Faculty of Agriculture, Hokkaido (Imperial) University--*J. Fac. Agric. Hokkaido (Imp.) Univ.*

Journal of Japanese Botany, Tokyo, 1916 - *.J. Jap. Bot.*

Journal of Mycology 1885-1908. J. Mycol. see pg. 104.

Journal of Natural Products, 1979-). Replaces *Lloydia*, see below.

Journal of the Tennessee Academy of Sciences, 1926- . *J. Tenn. Acad. Sci.*

Karstenia, 1950- . unchanged, see pg. 102.

Kawaka, 1973- . unchanged, see pg. 103.

Kew Bulletin, 1946 - *Kew Bull.* Also includes additional (Add.) Series see pg. 110.

Life Sciences and Agricultural Experimental Station: Bulletin University of Maine at Orono. Occasional papers.

Lloydia, 1938-1978. Unchanged. Now replaced by Journal of Natural Products; see above.

Magyar Gombaszati Lapok, 1944 - *Magy. Gomb. Lap.* see pg. 102.

Madrono, 1916 - unchanged.

Mededelingen Nederlandische Mycologische vereeniging. *Meded. ned. mycol. Vereae.* (1933-1951) see pg. 103.

Meddelelser om Grnland, Copenhagen, *Medd. Grn.*.

Memoirs of the Brooklyn Botanic Garden, 1918-1936. *Mem. Brooklyn Bot. Gdn.*

Memoirs of the New York Botanic Gardens, 1900 - *Mem. N.Y. Bot. Gdn.*

Memoirs of the Torrey Botanical Club, 1889 - *Mem. Torrey Bot. Club.*

Michigan Botanist, 1962 - *Mich. Bot.*

Mikologija i Fitopatologija, 1967 - *Mikol.* and *Fitopatol.*

Mycopathologia and Mycologia Applicata, 1938-1974;*Mycopath. Myc. Appl.* now replaced by Mycopathologia from Vol. 55, 1975-

Mycologia, 1909- Unchanged, see pg. 104. Includes Mycologia Memoirs, see pg. 104.

Mycological Papers, C.M. Inst. see pg. 104.

Mycotaxon, 1974- . Unchanged, see pg. 104.

Naturalist. Hull, London, 1864- . *Naturalist* (Hull). See pg. 104.

New Zealand Journal of Bottany, 1963-*N.Z.J.l Bot.*.

Northwest Science, 1927 - *N.W. Sci.*

Notes from the Royal Botanic Garden, Edinburgh, 1900- Notes *R.*

Bot. Gdn. Edinb. See pg. 110.

Nova Hedwigia, 1959-   Includes Beihefte and Icones Coloratae as supplement; abb. *Nova. Hedw.* Commercially produced journal by J. Cramer, Vaduz.

Nytt Magasin for Botanikk, 1952 - *Nytt Mag. Bot..*

Papers from the Michigan Academy of Science, Arts and Letters, 1921-   *Pap. Mich. Acad. Sci.*

(Die) Pilze Mitteleuropa. Series of iconographic works so far covering boletes, *Russula, Lactarius* and *Phlegmacium*, 1926-1967.

Phillipines Journal of Science, 1906 - *Philipp. J. Sci.*

Proceedings of the American Philosophical Society, 1838 -. *Proc. Am. Phil. Soc.*

Proceedings and Transactions of the Nova Scotian Institute of Science. 1863 - *Proc. (Trans.) Nova Scotian Inst. Sci.*

Prodome Flore Mycologique de Madagascar. A series of publications on the more conspicuous fungi of Madagascar. abb. *Prod. Fl. Mycol. Madagascar.*

Proceedings of the Koninklijke Nederlandse Akadamie van Wetenschappen, 1899 - *Proc. K. ned. Akadwet.*

Publicacions de l'Institut botanic, Barcelona. *Publcions Inst. Bot. Barcelona.*

Report of the Tottori Mycological Institute, Japan, 1963. *Rept. Tottori Myc. Inst.*

Results Norwegian Scientific Expedition to Tristan da Cunha, (1937-38), 1946 - *Results Norw. Scient. Exped. Tristan da Cunha.*

Revue de Mycologique, 1936-   . includes Mem. Hors. *Rev*(ue) *Mycol.* See pg. 102.

Rhodora, 1899 - unchanged.

Sbornik narodniho musea v Praze, 1938-   . *Sb. nar. Mus. Praze.*

Schweizerische Zeitschrift fur Pilzkunds, 1923 - *Schweiz Z. Pilzk.*

Soverskaya Botaniska, 1933 - *Sov. Bot..*

Sydowia, 1947-   . unchanged, includes Beihefte, see pg. 101.

Technical Communication of the National Botanic Garden, Lucknow, 1968. *Tech. Comm., Nat. Bot. Lucknow.*

Technical Papers. Division of Plant Industry, Commonwealth Scientific and Industrial Research Organization, Australia. *Tech. Pap. Div. Plant Ind. C.S.I.R.O. Aust.*

Technical Publications, State University College of Forestry at Syracuse Univ.; occasional papers.

Transactions of the British Mycological Society, 1897 - *Trans. Brit. Mycol. Soc..*

Transactions of the Botanical Society of Edinburgh, 1839 - *Trans. Bot. Soc. Edin.* See pg. 110.

Transactions of the Mycological Society of Japan. 1960 - *Trans. Mycol. Soc. Japan.*

Transactions of the Royal Society of South Australia, 1877 - *Trans. R. Soc. S. Aust.*

Transactions of the Wisconsin Academy of Science, Arts and Letters, 1870 - *Trans. Wis. Acad. Sci. Arts Letts.*

Travaux Scientificques du Parc National de la Vanoise, 1970 . *Trav. Sc. Parc Nat. Vanoise.*

Treballs Museu de ciencies naturals. see *Mus. barcin. Scient. nat. op.* 1917-1955.

Trudy Botanicheskogo Instituta. Akademiy Nauk SSSR, 1933 -.*Trudy Bot. Inst. Akad. Nauk SSSR.*

Zeitschrift fur Pilzkunde, 1922 - *Z. Pilsk.* Now Deutsche Gesellschrift fur Mykologie: Zeitschrift fur Mycologie. 1977 -. Now replaced by *Zeits. Mycol.* See pg. 103.

OCCASIONAL PAPERS

Europe:

Arkives for Botanik, 1950 - *Ark. Bot.* See pg. 92.

Archives du Museum National Histoire Naturelle, Paris, 1932. *Arch. Mus. Natn. Hist. Nat.* See pg. 62.

Boissera, 1936 - unchanged.

Bulletin de la Societe Botanique de Nord France 1947 - *Bull. Soc. Bot. Nord Fr.* See pg. 75.

Lietuvos TSR Mosklu Akademie darbai. see pg. 49.

Publique Par la Commission de la Societe Helvetique des Sciences naturelles pour les etudes scientifique du Parc National. See pg. 77.

Museum de Siencies Natural de Barcelona. *Mus. Siencies Nat. Barcelona.* See pg. 78.

Svensk Botanisk Tidskrift, 1907 - *Svensk bot. Tidskr.* See pg. 70.

Travaux du Laboratoire de La Jaysinia a Samoens (Haute-Savoie). Fondation Cognacq-Jay. 1969 - . Trav. Lab. Jaysinia, see pg. 60.

Travaux dedies a Viennot - Bourgin, see pg. 25. Trav. . .

Travaux Scientifiques du Parc national de la Vanoise. - Trav̇. so. parc nat. Vanoise, see pg. 24-25.

Japan

Annual Report of the Institute for Fermentation, Osako, Hakko Kenkyujo Nempo. see pg. 88.

Memoirs of the Shiga University Faculty of Arts Science. *Mem. Shiga Univ.* See pg. 81.

Memoirs of the Natural Science Museum, Tokyo, 1957. Mem. nat. Sci. Mus. Tokyo.

India:

Current Science, 1932 - . *Curr. Sci.* See pg. 62.

S.E. Asia:

Garden's Bulletin, Singapore, 1947 - *Gdn's Bull. Singapore.* See pgs. 14, 68.

S. Africa

Contributions from the Bolus Herbarium, S. Africa, 1969 -- . *Contr. Bolus Herb.* See pg. 85.

Central and South America

Anales de la Escuela nacional de Ciencias 6: Ologicas, 1948 --. *An. Elc. nac. Cienc. Biol. Mex.* See pg. 89.

Anales del Instituto de biologia. Universidad de Mexico 1930 - *An. Inst. Biol. Univ. (Auton.) Mex. Ser. Bot.*

*Anais da Sociedade de biologica de Pernambuco, 1938. *Ann. Soc. Biol. Pernamb.* See pg. 91.

Boletin de la Sociedad Botanic de Mexico, 1962 - *Bol. Soc. Bot. Mex.* See pg. 89.

*Contributions from the Institute Antarctic Argentina. See pg. 92.

*IMUR Acta - see Sydowia, Beih. 8 (1979). *IMUR Acta.* See pg. 91.

Vellozia 1961 - unchanged. See pg. 90.

*All articles quoted in this paper have been seen except those marked with a *.

C. INTERNATIONAL JOURNALS CARRYING A HIGH PROPORTION OF MYCOLOGICAL ARTICLES, ALTHOUGH NOT CONFINED TO THIS SUBJECT AREA.

*Botanik Tidss*krift. See below Dansk Bot. Arkiv.

*Bull*etin du *Jard*in *bot*anique *nat*ional de *Belg*ique (Bull. van de National Plntentium van Belgie) A. Robyns, Domaine de Bouchout B-1860, Meise, Belgium.

*Bull*etin (mensuel) de la *Societe Linn*eene de *Lyon*. 33, rue Bossuet 69006, Lyon, France.

*Can*adian Journal of *Bot*any, National Res. Council of Canada, Ottawa, KIA OR6, Canada.

*Dansk Bot*anik *Ark*iv. Dansk Bot. Forening Farmimagsgade 2 Dopg. G Dk-1353, Copenhagen, K, Denmark.

Israel Journal of *Bot*any, Weizmann Sc. Press of Israel, P.O. Box 801 Jerusalem, 91000, Israel.

Journal of *Nat*ural *Products, American Society of Pharmacognosy. Spahr and Glenn Co., 225 E. Spring St., Ohio, 43215, U.S.A.*

*Kew Bull*etin (continuation of Miscellaneous Papers, Kew) 1958 started Vol. 13 Official organ of the Royal Botanic Gardens, Kew, Richmond, Surrey, England.

(The) *Naturalist* (London Hull). Official organ of the Yorkshire Naturalists Union. D. Bramley & Doncaster Museum, Chequer Road, DN1 2AE Doncaster, England.

New Zealand Journal of *Bot*any, Scientific Information Officer, Dept. Scientific and Industrial Research, P.O. Box 9741, Wellington, New Zealand.

*Norw*eigan Journal of *Bot*any (formerly Nytt Magasin for Botanik up to Vol. 17, 1970). H. Krog, Bot. Museum, Trondheimsun 23B Oslo, Norway.

*N*otes from the *Royal Bot*anic *Garden*, Edinburgh. Official organ of the Royal Botanic Garden, Edinburgh, Scotland.

*Sym*bolae *Bot*anicae *Upsal*iensis (Acta Universitatis Upsaliensis). Institute of Systematic Botany Univ. of Uppsala, Sweden.

*Trans*actions (and Proceedings) of the *Bot*anical *Soc*iety of *Edin*burgh. Official organ of the Botanical Society of Edinburgh, c/o Royal Botanic Gardens, Inverleith Row, Edinburgh, Scotland.

INDEX

112

114

118

APPENDIX
ATLAS DE LA SOCIETE MYCOLOGIQUE DE FRANCE: PLATE NOS.

Agaricus rubellus	22, Vol. 43 (1927)
Agaricus spp. see Psalliota below	
Amanita codinae	186 Vol. 87 (1971)
A. eliae	46 Vol. 46 (1930)
A. umbrinolutea	29 Vol. 45 (1929)
A. vittadini	186 Vol. 87 (1971)
Boletus armeniacus	190 Vol. 98 (1972)
B. collinitus	174 Vol. 83 (1967)
B. corsicus	189 Vol. 87 (1971)
B. dupainii	44 Vol. 47 (1931)
B. fragrans	45 Vol. 47 (1931)
	& 150 Vol. 81 (1965)
B impolitus	46 Vol. 47 (1931)
B. lepideus	147 Vol. 80 (1964)
B. leucophaeus	126 Vol. 77 (1961)
B. lupinus	90 Vol. 64 (1948)
B. luteoporus	189 Vol. 87 (1971)
B. niveus?	127 Vol. 77 (1960)
B. oxydabilis	127 Vol. 77 (1960)
B. purpureus	12 Vol. 42 (1926)
B. queletii	12 Vol. 42 (1926)
	& 66 & 67 Vol. 51 (1935)
B. regius	17 Vol. 42 (1926)
B. regius subsp. torosus	18 Vol. 42 (1926)
B. subtomentosus agg.	18 Vol. 42 (1926)
	& 42 & 43 Vol. 47 (1931)
B. tridentinus	89 Vol. 63 (1947)
B. viscidus	91 Vol. 65 (1949)
Boletus, see also Ixocomus, Leccinum & Xerocomus	
Cantharellus melanoxeros	199 Vol. 91 (1975)
Clitocybe gallinacea	64 Vol. 50 (1934)
C. martiorum	143 Vol. 80 (1964)
C. martiorum	143 Vol. 80 (1964)
C. tabescens	55 Vol. 49 (1933)
Collybia tylicolor = Tephrocybe	80 Vol. 54 (1938)
Coprinus insignis	23 Vol. 44 (1928)
Cortinarius, see separate list	
Deconica atrorufa = Psilocybe	73 Vol. 52 (1936)
Drosophila silvestris = Psathrella	106 Vol. 71 (1955)
Flammula carbonaria = Pholiota	31 Vol. 45 (1929)
Galera lateritia = Conocybe lactea	56 Vol. 49 (1933)
Gyrodon lividus = Uloporus	94 Vol. 66 (1950)
Hygrophorus arbustivus	162 Vol. 82 (1966)
H. colemannianus	163 Vol. 82 (1966)
H. foetens	6 Vol. 41 (1925)

H. lacmus	11 Vol. 46 (1930)
H. limacinus	180 Vol. 85 (1970)
H. laetus	37 Vol. 45 (1929)
H. penarius	117 Vol. 75 (1959)
H. poetarum	102 Vol. 70 (1954)
Hypholoma dispersum	140 Vol. 80 (1964)
Inocybe atripes	105 Vol. 71 (1955)
I. fibrosa	169 Vol. 83 (1967)
I. globocystis	54 Vol. 48 (1932)
I. napipes	53 Vol. 48 (1932)
I. sambucina	111 Vol. 73 (1957)
Ixocomus lignicola = Boletus	154 Vol. 81 (1965)
Lactarius acris	144 Vol. 80 (1964)
L. decipiens var. lacunarum	82 Vol. 55 (1939)
L. evosmus	179 Vol. 84 (1968)
L. griseus	27 Vol. 44 (1928)
L. mitissimus	84 Vol. 57 (1941)
L. representaneus	81 Vol. 55 (1939)
L. ruginosus	145 Vol. 80 (1964)
L. speciosus	146 Vol. 80 (1964)
	& 179 Vol. 84 (1968)
L. spinulosus	27 Vol. 44 (1928)
L. subdulcis	85 Vol. 57 (1941)
Leccinum duriusculum	184 Vol. 86 (1970)
Lentinus degener	170 Vol. 83 (1967)
L. gallicus	30 Vol. 45 (1929)
L. tigrinus	178 Vol. 84 (1968)
L. variabilis	25 Vol. 44 (1928)
Lepiota badhami	1 Vol. 41 (1925)
L. bucknalii	72 Vol. 52 (1936)
L. castanea	74 Vol. 53 (1937)
L. cretini	71 Vol. 52 (1936)
L. echinata = Melanophyllum	72 Vol. 52 (1936)
L. fulvella	74 Vol. 53 (1937)
L. laevigata	71 Vol. 52 (1936)
L. ochraceofulva	172 Vol. 83 (1967)
Leucopaxillus amarus	24 Vol. 44 (1928)
L. lepistoides	187 Vol. 87 (1971)
Limacella megalopoda	15 Vol. 42 (1926)
L. gliocyclum	20 Vol. 43 (1927)
Lyophyllum leucophaetum	87 Vol. 59 (1943)
Mycena flavipes	96 Vol. 66 (1950)
M. osmundicola	70 Vol. 52 (1936)
M. seynii	96 Vol. 66 (1950)
Naucoria christinae = Phaeocollybia	38 Vol. 46 (1930)
N. lugubris =Phaeocollybia	39 Vol. 46 (1930)
Omphali(n)a asterospora	110 Vol. 73 (1957)
O. chrsophylla	28 Vol. 44 (1938)
O. griseo-pallida	80 Vol. 54 (1938)

Cortinarius
C. alborufescens	103 & 104 Vol. 71 (1955)
C. allutus	124 Vol. 77 (1961)
C. aremoricus	139 Vol. 79 (1963)
C. balteatoalbus	205 Vol. 93 (1977)
C. boudieri	113 Vol. 73 (1957)
C. bulbosus	132 Vol. 79 (1963)
C. callisteus	151 Vol. 81 (1965)
C. claroflavus	201 Vol. 92 (1976)
C. cyanobasalis	206 Vol. 93 (1977)
C. depressus	89 Vol. 54 (1938)
C. diabaphus	101 Vol. 70 (1954)
C. eufulmineus	219 Vol. 94 (1978)
C. evosmus	123 Vol. 77 (1961)
C. flavovirens	121 Vol. 76 (1960)
	& 202 Vol. 92 (1976)
C. fulvoochraceus	201 Vol. 92 (1976)
	& 202 Vol. 92 (1976)
C. herbarum	125 Vol. 77 (1961)
C. herculeus	177 Vol. 84 (1962)
C. limonius	129 Vol. 78 (1962)
C. lividoviolaceus	185 Vol. 79 (1963)
C. luteoiimarginatus	118 Vol. 75 (1959)
C. lutescens	116 Vol. 74 (1958)
C. mucifluoides	136-138 Vol. 79 (1963)
C. multiformis	124 Vol. 77 (1961)
C. ochropallidus	122 Vol. 77 (1961)
C. orellanus	128 Vol. 78 (1962)
C. parevernius	134 Vol. 79 (1963)
C. polymorphus	125 Vol. 77 (1961)
C. praesignis	133 Vol. 89 (1963)
C. prasinocyaneus	203 Vol. 93 (1977)
C. prasinus	119 Vol. 76 (1960)
C. pseudosuillus	207 Vol. 93 (1977)
C. rapaceus	122 Vol. 77 (1961)
C. rheubarbarinus	132 Vol. 79 (1963)
C. speciossissimus	153 Vol. 81 (1965)
C. suaveolens	83 Vol. 56 (1940)
	& 148 Vol. 81 (1965)
C. subionochlorus	149 Vol. 81 (1965)
C. subrachodes	204 Vol. 93 (1977)
C. uliginosus	152 Vol. 81 (1965)